Extraction of Semiconductor Diode Parameters

Richard Ocaya

Extraction of Semiconductor Diode Parameters

A Comparative Review of Methods
and Materials

 Springer

Richard Ocaya
Department of Physics
University of the Free State
Bloemfontein, South Africa

ISBN 978-3-031-48849-8   ISBN 978-3-031-48847-4 (eBook)
https://doi.org/10.1007/978-3-031-48847-4

This Springer imprint is published by the registered company Springer Nature Switzerland AG
The registered company address is: Gewerbestrasse 11, 6330 Cham, Switzerland

Paper in this product is recyclable.

... to the children, for your uplifting spirits ...

Foreword by Prof. Dr. Fahrettin Yakuphanoğlu

This monologue serves as a fitting introduction to the critical parameters of semiconductor diodes. It is precisely a monologue for its insistence on conveying the importance of key parameters, namely series resistance, ideality factor, barrier height, and temperature, which encompass both intrinsic and extrinsic aspects.

The motivation behind this specific task is clearly not to regurgitate what has come before, much of which is already pristine and exemplary. Instead, the main theme is to highlight what has become contemporary such that even newcomers will quickly become abreast of the essentials. Therefore, it hopes to have a positive impact on a new generation of researchers. In approaching the task, the key parameters are kept squarely in sharp focus throughout the entire work. Furthermore, what sets this work apart from others in the field is its comprehensive three-pronged approach: introduce, define, and refer.

The author begins by acknowledging the importance of accurately representing prior research focused on parameter extraction for both p-n and metal-semiconductor diodes. The author, whom I have had the privilege of knowing for two decades, has been a stalwart in this field, making significant contributions both independently and through global collaborations.

With this extensive experience, the author is well-positioned to identify emerging innovations that can invigorate and propel the field of parameter extraction forward. The chosen topics align with the current concerns of the community, particularly in characterizing the increasingly miniaturized devices, ultimately aiming to provide more robust tools for elucidating the underlying current transport mechanisms in these devices.

One highlight of this monologue is the chapter on the potential unification of current-voltage-temperature measurements by addressing an underlying symmetry. While the method may appear formidable at first glance, the rationale and final outcomes elegantly embody simplicity.

Furthermore, those abreast of the latest developments in computing will undoubtedly recognize the significance of the chapter on the exponential growth of artificial intelligence methods in data science, with profound implications for the physical sciences. The discussion on AI techniques underscores the potential for automation and expedited device development using this burgeoning tool, which is poised for continuous expansion.

In sum, this book presents a much-needed and focused contribution that will resonate with graduate students, researchers, and educators engaged in the realm of semiconductor device fabrication and characterization.

Elaziğ, Turkiye   Prof. Dr. Fahrettin Yakuphanoğlu
September 2023

Foreword by Kwadwo Mensah-Darkwa

General Overview

The book addresses the rapidly evolving field of semiconductor device characterization, which necessitates a thorough and up-to-date resource covering a wide range of device types and materials. The book acknowledges that while it is challenging to keep up with the daily progress in this field, its goal is to review recent advances in metal-semiconductor, metal-insulator semiconductor, and p-n junction diodes, which utilize various semiconductor materials, including elemental, compound, organic, and nanostructured materials. The book includes fundamental knowledge of the extraction of characteristic parameters of these devices.

One key strength of this book is its commitment to providing a comparative analysis of methods used to extract semiconductor diode parameters, especially under the influence of temperature.This unique feature aims to equip readers with the knowledge to make informed decisions about selecting the most suitable method for their specific needs.

The book also emphasizes the importance of establishing a standardized point of reference within the field, particularly regarding the internal parameters of metal-semiconductor junction devices.

Furthermore, the book focuses on critical device parameters that have yet to receive sufficient attention in the existing literature and explains how this focus is interwoven with other parameters in semiconductor device characterization.

Market Value

This book will prove highly beneficial for students, scientists, and researchers specializing in the fields of metal-semiconductor, metal-insulator-semiconductor, and p-n junction diodes.

The book covers a wide range of essential topics, including the determination of p-n junction band gaps, a comprehensive review of Metal-Semiconductor Junctions, contemporary Parameter Extraction Methods, and more. These topics make the book suitable for graduate-level courses and can be an introductory resource for undergraduate classes in the same subject areas.

Additionally, the book delves into emerging subjects like artificial intelligence and machine learning in the context of parameter extraction methods.

I recommend this book to my postgraduate students and suggest it be added to our university library's collection.

Conclusion

In summary, this book aims to be a valuable and indispensable resource for researchers, engineers,and enthusiasts in the field of semiconductor device characterization, with a focus on empowering them with the knowledge and tools needed to advance the understanding of semiconductor devices and contribute to technological innovations.

Regards,

January 2024

Kwadwo Mensah-Darkwa, Ph.D., MGhIE
Commonwealth Academic Fellow
Senior Lecturer
Faculty of Mechanical and Chemical Engineering
Department of Materials Engineering
College of Engineering
Kwame Nkrumah University of Science and Technology
Kumasi, Ghana
kmensahd@aggies.ncat.edu

Preface

The rapidly advancing field of semiconductor device characterization demands a comprehensive and authoritative update of resources that addresses the diverse developments around the device type and materials. While a single source cannot realistically hope to fulfill this mission to perfection in the light of literally daily progress, this book aims to encapsulate the recent advances in the exploration of metal-semiconductor, metal-insulator-semiconductor, and p-n junction diodes. These devices employ a wide span of semiconductors encompassing elemental, compound, organic, and nanostructured materials. Such comprehensive coverage is notably absent from existing literature, which tends to concentrate on only specific diodes or materials, leaving readers craving a more holistic understanding.

An exceptional aspect of this book is its comparative analysis of different methods employed to extract semiconductor diode parameters, particularly under the influence of temperature. This unique feature is seldom found in existing literature, offering readers an invaluable opportunity to discern the strengths and weaknesses of each approach. By equipping readers with this knowledge, the book empowers them to select the most suitable method for their specific requirements, fostering a deeper understanding and enabling more informed decision-making.

Crucially, this book aims to establish a standardized point of reference within the field, addressing a significant gap in the current landscape, with the specific scope of determining the internal parameters of metal-semiconductor junction devices, starting from the complex thermionic emission model. In the absence of a recent, single resource that provides a comprehensive and up-to-date review of application-driven methods for extracting semiconductor diode parameters, researchers and engineers often find themselves navigating a fragmented information landscape. This resource hopes to alleviate this challenge, providing a much-needed centralized reference that researchers and engineers can rely on for efficient information. This book is relevant because its central focus is the series resistance, a key parameter in deviating from the popular thermionic emission model. The reasons for this sharp focus become evident in the early chapters, as series resistance intricately interweaves the other parameters. Yet, despite its importance, up to this point, its treatment in the literature is cursory.

In the following pages, readers will journey through the intricate world of semiconductor device characterization, guided by a meticulous focus on the crucial parameters of p-n and MS diodes, materials, and the present extraction methods. Furthermore, the book delves into emerging 2D materials and their hetero-structures in real devices. The book highlights clearly the need for an advancement in the physics of their operation in contrast to the traditional approaches of their 3D counterparts. Today, with the rise of artificial intelligence (AI), no work can be considered complete without even cursorily considering the potential of the emerging technology. We dedicate a chapter to this emerging field and show how appropriate AI models can learn parameter extraction without knowing the underlying physics. This statement will alarm many, but it is an inevitable debate that must happen somewhere else.

Finally, this book is a testament to the author's continued presence in the field and a commitment to empowering researchers, engineers, and enthusiasts with the knowledge and tools needed to unlock the full potential of new semiconductor devices. We hope this resource will become an indispensable companion, propelling the field forward and inspiring groundbreaking innovations that shape the future of technology.

Qwaqwa, Republic of South Africa   Richard Ocaya
June 2023

Acknowledgements

Creating a resource such as this, that aspires to encapsulate the latest and most impactful advancements in the field of semiconductor measurements, is an immense and, possibly, foolhardy endeavor. What is most certain is that bringing it to fruition such a work is a monumental task requiring the collective efforts of numerous dedicated individuals. Thus, I wholeheartedly extend my gratitude to all these contributors, and I genuinely express my regret if any contribution has unintentionally been omitted.

Most notably, I acknowledge Prof. Dr. Fahrettin Yakuphanoğlu of Firat University, Turkiye, to whom I am eternally grateful for the invitation to join his stellar research group over a decade ago, and then subsequently for the friendship that continues to blossom to this day. Our many collaborations, which span a wide range of optical and thermal sensing metal-semiconductor devices, have been eye-openers by revealing the promise of Group II–VI elements with p-type silicon to solve some of humanity's contemporary problems. Specifically, I acknowledge and appreciate the culmination of our discussions on a novel method for Schottky diode parameter extraction that injected fresh breath into the field in almost half a century.

I also wish to thank my past and present colleagues and students at the Department of Physics at the University of the Free State (UFS) for the innumerable interactions over the years. These encounters refined my understanding of the concepts now presented in this work. Furthermore, I appreciate the supportive atmosphere possible in the department through your conscious efforts. I convey a special gratitude to Dr. Kamohelo G. Tshabalala for being the primary dynamo of this department.

I wish to convey special thanks to Dr. Steven Rwabona Katashaya for being a great friend and valued advisor over the last two decades.

Last but not the least, I extend my sincere gratitude and appreciation to Ms. Nthabiseng Moloi at the UFS, for her tireless preparation of the text and diagrams in the manuscript.

Contents

Acronyms

The following abbreviations and symbols are used in this book.

MS	Metal-semiconductor
MSJ	Metal-semiconductor junction
TE	Thermionic emission
φ	Work function
Φ	Potential energy barrier height
Φ_0	Flat-band potential energy barrier height
eV	Electron volt, equal to a joule per electron charge
XPS	X-ray photoelectron spectroscopy
DOS	Density of states of energy
D_{it}	Density of interface states on the surface of a semiconductor
N_{ss}	Is the same as D_{it}
χ	Electron affinity
FD	Fermi-Dirac probability distribution
BHI	Barrier height inhomogeneity
E_F	Fermi level
w.r.t	With respect to
s.t.	such that
2D, 3D	Two-dimensional and three-dimensional
ODE	Ordinary Differential Equation
LSC	Linearized symmetry condition
iff	If-and-only-if
ML	Machine Learning
DLTS	Deep-level transient spectroscopy

Chapter 1
Determining p-n Junction Band Gap

Abstract The present chapter describes a new technique for the accurate and reliable determination of p-n diode bandgap energy. The method makes no assumptions about the ideality factor and therefore differs from many experimental techniques. The well-known linear variation of p-n diode terminal voltage with temperature at fixed forward current is still used, but it is shown that if the current through the diode is switched in controlled steps, then it is possible to determine the temperature difference needed to maintain a constant diode terminal voltage. The ideality factor and bandgap can then be accurately determined over a small temperature range at well-defined voltage-temperature points, termed the measurement trajectory. An application is presented using a carefully programmed microcontroller circuit to determine the temperature trajectory for the 1N4148 silicon and OA91 germanium diodes. The microcontroller is being used increasingly in university science laboratories for its availability and simplicity of use [1]. The results show good agreement with the literature values for silicon bandgap energy but give a 30% higher value for germanium.

1.1 Importance of Temperature

The current I flowing through a p-n diode at applied voltage V and junction temperature T is given by Shockley's equation:

$$I = I_n \exp\left(-\frac{E_g}{nkT}\right)[\exp(qV/nkT) - 1]. \qquad (1.1)$$

where I_n is a constant, k is Boltzmann's constant, q is electronic charge, n is a diode-dependent ideality factor, and T is absolute temperature. E_g is the bandgap of the semiconductor [2–6]. Under the condition that $qV/nkT >> 1$, Eq. 1.1 is approximately

$$I \approx I_n \exp\left(-\frac{E_g}{nkT}\right)\exp\left(\frac{qV}{nkT}\right). \qquad (1.2)$$

It is easier in practice to fix the current through the diode and measure the terminal voltage, according to the equation

© The Author(s), under exclusive license to Springer Nature Switzerland AG 2024
R. Ocaya, *Extraction of Semiconductor Diode Parameters*,
https://doi.org/10.1007/978-3-031-48847-4_1

$$V(T) = \frac{E_g(0)}{q} + \frac{nkT}{q} \ln\left(\frac{I}{aT^b}\right), \tag{1.3}$$

where a and b are constants [7]. The p-n diode current clearly also has a strong temperature dependence that makes it suitable as a thermal sensor [5, 7]. The logarithmic term in Eq. 1.3 is much less sensitive to temperature change than the linear T term [7, 8], and this gives a thermal sensor that is linear from cryogenic temperatures up to about 200 °C. The temperature sensitivity varies between 1 and 3 mV/°C, which, however, has been shown to have a slight dependence on the diode current [4]. In practice, the ideality factor is significant to the properties of the diode and is due to a combination of factors within the semiconductor, such as surface effects, recombination, and tunneling.

More often than not, the ideality factor has been assumed as unity prior to the experiment [1, 9]. Many experiments have shown that a non-unity ideality factor is more likely to be found in practical diodes [3, 5]. The technique described here solves this problem. The ideality factors reported in the literature range from 1.4 to 2.0 for silicon and 1.7 to 2.2 for germanium [10]. Despite being well-studied, the humble p-n junction diode remains an important device in the solid-state physics laboratory because it provides a low-cost way to illustrate important concepts such as bandgap, barrier potential, reverse saturation current, and the sensitivity of semiconductors to thermal energy. Bandgap does vary with temperature, and a number of empirical determinations of this variation exist [3, 9, 11]. For example,

$$E_g(T) = E_g(0) - \frac{\alpha T^2}{\beta + T}, \tag{1.4}$$

where $E_g(0) = 1.17\,\text{eV}$ is 0 K bandgap, $\alpha = 0.473\,\text{meV/K}$, and $\beta = 636\,\text{K}$ [3]. Another advantage of the present approach is that it establishes a characteristic temperature variation of only about 18 K. According to Eq. 1.4, it is easy to show that given a temperature uncertainty of ΔT, the uncertainty in the bandgap is

$$\Delta E_g \approx \pm \frac{2\alpha T}{\beta + T} \Delta T. \tag{1.5}$$

An uncertainty analysis using Eq. 1.5 for an experiment conducted around 332 K (60 °C) over an 18 K temperature range gives a bandgap of 1.168±0.001 eV. Therefore, the bandgap calculated using the present approach is an accurate representation of the bandgap at the center temperature. Impurities [12] and doping density [13] also affect bandgap, but for a given diode, these factors are beyond control in this experiment. The novel approach described below shows that if the constant forward current through the diode is switched in a controlled manner, it is possible to estimate n and E_g using only measurements of the temperatures at which the new constant current reasserts the set forward voltage as the temperature changes. The simplicity of this new approach makes it suitable for both manual and automated characteriza-

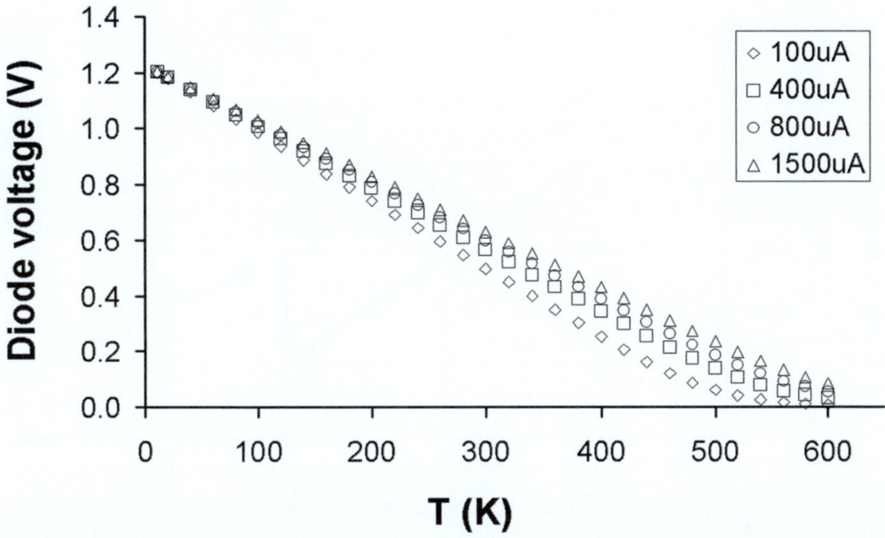

Fig. 1.1 Typical variation of forward diode voltage with temperature and current for the 1N4148 silicon diode. (Reproduced from [4])

tion of n and E_g.   Figure 1.1 shows the typical curves for the 1N4148 diode with constant currents below 4 mA, for which self-heating effects are negligible [4].

1.1.1   Theory

The main interest is to determine the bandgap using only two sets of measurements at determined temperature points. Consider Eq. 1.1 and Fig. 1.2. Suppose that the diode is heated to a high initial temperature T_0 beyond point A in the figure, whereupon the heat source is removed. Alternatively, rapidly boiling water may be used. With the forward current set to I_1, the diode then cools to T_1. The forward voltage of the diode is then measured as V_{F1} (point B in Fig. 1.1), and the diode current is then switched to a new constant value I_2, where $I_2 \leq I_1$. In the application, it was noticed that a short delay of about 500 ms was useful to allow the new current to stabilize before reading the new terminal voltage V_{F2} (at point C). As the diode continues to cool down under the new current I_2, its terminal voltage will eventually reach V_F again, at which the measured temperature will be T_2. These measurements are sufficient to determine the ideality factor and bandgap.

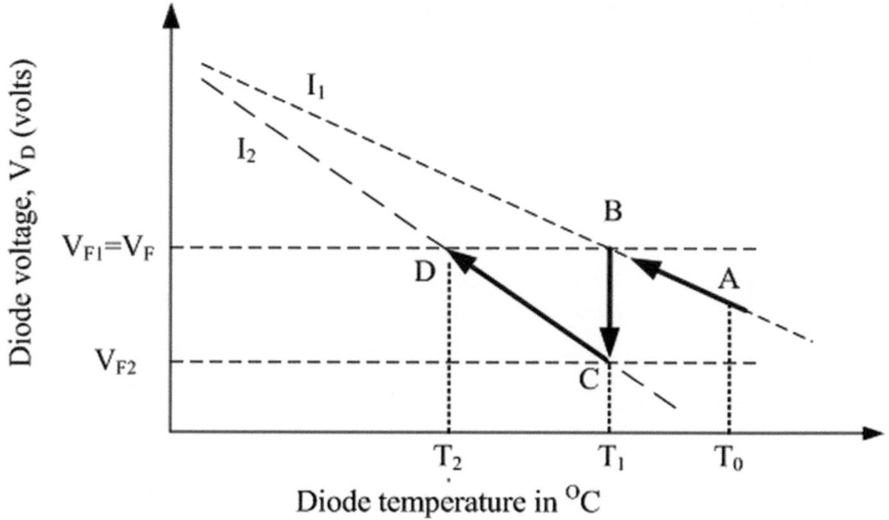

Fig. 1.2 Depiction of the cooling and measurement trajectory (solid lines) under constant voltage and currents

1.1.1.1 Determining the Ideality Factor

The constants C_1, C_2, and $E_g(0)$ in Eq. 1.3 can be avoided by using two bias currents on the same device sequentially or on two identical devices simultaneously [7]. At a temperature T, the use of two sequentially switched currents gives the data pair: (T, I_1, V_1) and (T, I_2, V_2). Substituting the data pair into Eq. 1.3 allows the difference voltage $\Delta V(T) = V_1(T) - V_2(T)$ to be readily found. Consequently, the ideality factor is easy to find, i.e.:

$$n = \frac{q\,\Delta V(T)}{kT\,\ln(I_1/I_2)}.$$

(1.6)

Now consider Eq. 1.6 in the context of Fig. 1.2. The ideality factor can be estimated in the part of the measurement trajectory labeled B→C, where the temperature is T_1, and $\Delta V(T_1) = V_{F1} - V_{F2}$.

1.1.1.2 Determining the Band Gap

The mathematical implications of the sequential constant current switching on bandgap calculation are as follows. Suppose that the two currents are switched in a known ratio a, such that

$$I_2 = aI_1,$$

where $0 < a < 1$. It follows that for any two temperature endpoints at which the diode voltage is the same (i.e., V_F), Eq. 1.1 can be written

$$I_1 \approx I_n \exp\left(-\frac{E_g}{nkT_1}\right) \exp\left(\frac{qV_F}{nkT_1}\right). \tag{1.7}$$

and

$$aI_1 \approx I_n \exp\left(-\frac{E_g}{nkT_2}\right) \exp\left(\frac{qV_F}{nkT_2}\right). \tag{1.8}$$

Dividing Eqs. (1.7) and (1.8) and defining a new temperature T^* such that

$$\frac{1}{T^*} = \frac{1}{T_2} - \frac{1}{T_1}, \qquad \text{or} \qquad T^* = \frac{T_1 T_2}{T_1 - T_2}. \tag{1.9}$$

leads to

$$E_g = qV_F - nkT^* \ln a \qquad \text{(in joules, J)} \tag{1.10}$$

or

$$E_g = V_F - \frac{nkT^*}{q} \ln a \qquad \text{(in electron-volts, eV)}. \tag{1.11}$$

The band gap can be estimated in the part of the measurement trajectory labeled C→D, where a characteristic temperature T^* can be determined.

1.1.1.3 Bandgap in Terms of Potential and Temperature Ratio

An interesting implication of the above equations arises when Eq. 5.11 (with $T = T_1$) is substituted into Eq. 1.11. That is

$$E_g = V_F - \left[\frac{-q\Delta V(T_1)}{kT_1 \ln a}\right] \frac{kT^* \ln a}{q}. \tag{1.12}$$

Substituting T^* from Eq. 1.9 then gives the result

$$E_g = V_F + \frac{T_2}{T_1 - T_2} \Delta V(T_1). \tag{1.13}$$

This result achieves two things. Firstly, it eliminates the universal constants from the determination of bandgap. Secondly, it shows that bandgap can easily be determined by measuring the voltage change at two temperatures if the currents are sequentially switched at the two temperatures. This implies that it is not necessary to precisely know the values of the currents themselves, but simply to assure that the currents are repeatable. In short, Eqs. (5.11) and (1.11) may be used directly as the basis of the experimental measurement of ideality factor and bandgap. A mandatory requirement in the experiment is a selectable dual constant current source. Other requirements for measuring these parameters using an automated interface are accurate sensors for diode temperature, diode terminal voltage, reference voltage, reliable analog to digital data acquisition and timing, control, and readout software.

1.1.2 The Experimental Setup

The method used is the controlled cooling of an electrically insulated p-n diode in a water bath. The hot bath is constantly stirred as it cools down from near $100\,°C$. The diode temperature is sensed by an electrically insulated LM35CZ precision centigrade sensor that is connected to one channel of the PIC analog-to-digital converter (ADC). Electrical insulation of the devices that are immersed in the water bath was achieved by coating each device in non-conductive enamel paint that was then allowed to dry prior to the experiment.

1.1.2.1 Design of the Switchable Current Source

Figure 1.3 shows the design of the current source. It is an adaptation of the current mirror made up of matched Q1 and Q2 [5, 14]. An additional transistor Q3 switches the load on the collector circuit of Q1 depending on the state of its base current. The output current of the simple current mirror is given by

$$I = \frac{V_A}{R_C} \tag{1.14}$$

where R_C is the effective collector resistance of Q1, and the potential (V_A) at point A is nominally $(V_{CC} - V_{BE})$ above ground. When point C is at $0\,V$ (logic 0), Q3 is off and therefore the collector of Q1 has an effective resistance R_C equal to $(R_{C1}+R_{C2})$. When point C is at $+5V$ (logic 1), Q3 is driven into full conduction, and R_{C2} is effectively shorted to ground. Point B is therefore at $0V$ and the effective collector resistance is R_{C1}. In this way, the output current is programmable to

$$I_1 = \frac{V_{CC} - V_{BE}}{R_{C1} + R_{C2}}\bigg|_{C=0}, \quad \text{or} \quad I_2 = \frac{V_{CC} - V_{BE}}{R_{C1}}\bigg|_{C=1}. \tag{1.15}$$

Table 1.1 gives the transfer characteristic of the current source under digital control.

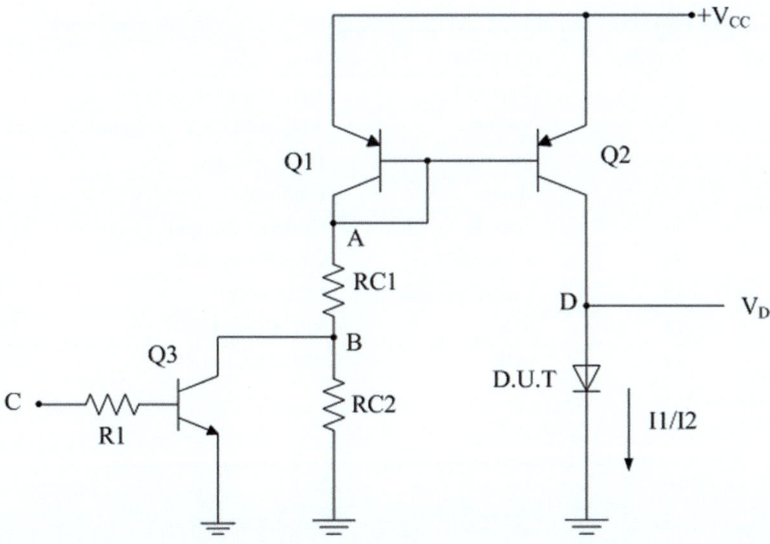

Fig. 1.3 A simple design of the I_1 or I_2 switchable current source controlled by a digital input at point C

Table 1.1 Transfer characteristics of the switchable current source

Point C (logic)	V_A (V)	V_B (V)	Output current (μA)	Name
0	4.33	2.3	106	I_2
1	4.31	0.0	228	I_1

1.1.2.2 Microcontroller-Based Measurement

The measurement can be automated in any number of ways. Here, a microcontroller is suggested. It is based on a pic16f877 microcontroller running at 4 MHz was used. The 10-bit ADC of the PIC was configured to run off the internal resistor-capacitor (RC) clock. Any PIC that supports at least five analog to digital input channels to a 10-bit ADC and at least eight output control lines can be used.

In this application, the readout scheme was simplified using a low-cost graphical liquid crystal display (LCD) unit. The PIC communicates with the LCD through the common serial peripheral interface (SPI) standard. However, the readout scheme used is a matter of choice and does not detract from the equations above. Table 1.2 summarizes the input/output lines used on the PIC.

Figure 1.4 shows the complete schematic diagram of the microcontroller interface.

Table 1.2 The pic16f877 signal lines and their designated function in this application

Pin name	Number	Function	Note
RA0	2	Vref $= V_{F1}$	AN0 (External calibration)
RA1	3	Vtemp	AN1 (LM35CZ centigrade sensor)
RA2	4	V_D	Diode voltage
RB0	33	Heater	(optional)
RB1	34	I_1 or I_2	Current selection
RB2	35	Stirrer motor	ON/OFF control
RC0	15	Data/$\overline{\text{Command}}$	LCD mode
RC3	18	SD0	SPI data to LCD
RC5	24	SCK	SPI clock to LCD

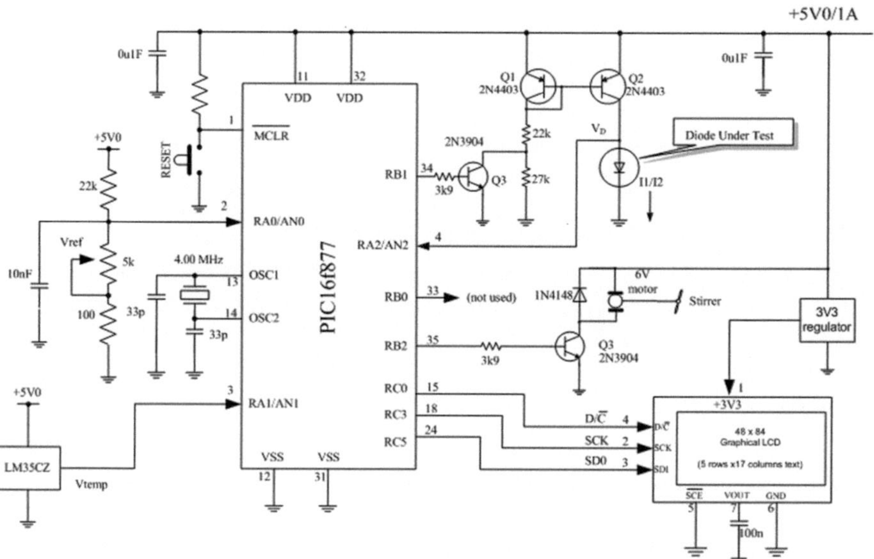

Fig. 1.4 Complete schematic diagram of the bandgap and ideality factor measurement circuit

1.1.2.3   The Programming Algorithm

Figure 1.5 shows the flowchart followed to compute the ideality factor and the bandgap of the diode under test (D.U.T) in Fig. 1.4. A high-level language such as C makes the process straightforward. The control program was coded in the C language using the mikroC Pro compiler.

The temperature of the water bath was monitored using the LM35CZ precision temperature sensor. The sensor has a sensitivity of 10 mV/°C. Conversion of its output to a kelvin temperature on the 10-bit ADC is done through the equation:

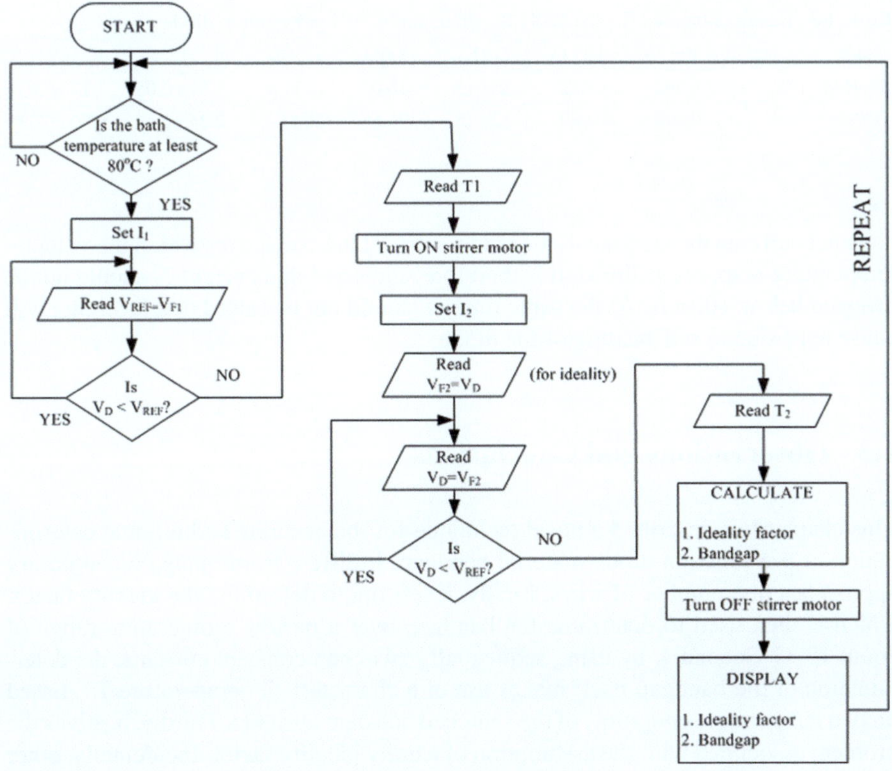

Fig. 1.5 Algorithm for determining ideality factor and bandgap

$$T(K) = \left[100V_M \times \frac{AN1}{1023} \right] + 273.15, \tag{1.16}$$

where AN1 is the decimal value of the ADC output converted from the LM35CZ (ADC channel 1), and V_M is the ADC reference voltage used (+5.0V in this case).

1.2 Results

Table 1.3 shows the experimentally determined ideality factors and bandgap using the setup with the 1N4148 silicon and OA91 germanium diodes respectively.

The uncertainties shown for the ideality factor and bandgap in Table 1.3 assume that all voltage and temperature measurements are equally subject to worst-case uncertainties of 0.01 V and 3 °C respectively. In practice, the measured accuracies tend to be much higher, especially when the microcontroller approach is used. In conducting the experiment with germanium diodes it should be noted that lower

Table 1.3 Results obtained for the 1N4148 silicon and OA91 germanium diode

Diode	Material	V_{F1} (V)	V_{F1} (V)	T_1 (K)	T_2 (K)	T^* (K)	n	E_g (eV)
1N4148	Si	0.462	0.422	333.15	316.05	6157.4	1.81±0.05	1.17±0.03
OA91	Ge	0.127	0.080	320.15	301.65	5220.2	2.24±0.07	0.90±0.03

constant currents through the diode tend to restrict the range over which the voltage-temperature response is linear. It is therefore suggested that current I_2 should not be dropped below $100\,\mu$A. At the same time, it should not be raised to a level that will cause appreciable self-heating of the diode.

1.3   Observations and Conclusions

This chapter has described a novel technique for the accurate and reliable determination of p-n junction diode material bandgap. It differs from many contemporary approaches in a number of ways. Firstly, it sets out to determine the ideality factor, which is then used to determine the bandgap over a narrow temperature range of about 18 °C. Secondly, by using sequentially switched constant currents, the determination of the bandgap itself makes use of a characteristic temperature T^*, based on two temperature endpoints of the switched constant currents. Thirdly, it solves the problem associated with the assumption of a unity ideality factor. Incidentally, other researchers have also shown that the ideality factor tends to be anything but unity. Ideality factor must therefore be considered to reach a more reliable estimate of the bandgap. Fourthly, the inherent limited temperature range also limits the spread of bandgap as a result of temperature as suggested by the uncertainty analysis using an empirical variation.

It is clear that only the temperature and voltage measurements pose any significant errors in the determined bandgaps. However, using precision temperature sensors and voltage measuring methods at the same endpoints cancels out these errors. Determinations of ideality factor and bandgap are therefore repeatable and reliable. Furthermore, by switching the currents in a known ratio, the equations ultimately show that precise knowledge of the current is unimportant, only that it is repeatable.

It is clear that making the switched current ratio small will reduce the temperature range of the measurement. However, an ADC should be chosen with care by paying attention to the measurement resolution and accuracy. While the experiment can be conducted manually with comparable ease, the use of the microcontroller eliminates the human component in the measurements and further increases the accuracy and reliability of the results. Finally, the values of the ideality factor of the 1N4148 diode (silicon) and the OA91 diodes (germanium) are in good agreement with the values reported in the literature. The determined value of the silicon bandgap energy agrees well with the literature, but was 30% higher for germanium, showing that ideality factors must not be assumed at unity.

References

1. R. Mukaro, B.M. Taele, D. Tinarwo, Europ. J. Phys. **27**, 531 (2006)
2. D.A. Fraser, *The Physics of Semiconductor Devices*. Oxford Physics Series, 4th edn. (Clarendon Press, Oxford, 1986)
3. B. Van Zeghbroeck, Principles of Semiconductor Devices. http://ece-www.colorado.edu/bart/book/
4. R.O. Ocaya, F.B. Dejene, Estimating p-n diode bulk parameters, bandgap energy and absolute zero by a simple experiment. Europ. J. Phys. **28**, 85 (2007). https://doi.org/10.1088/0143-0807/28/1/009
5. R.O. Ocaya, An experiment to profile the voltage, current and temperature behaviour of a P-N diode. Europ. J. Phys. **27**, 625 (2006). https://doi.org/10.1088/0143-0807/27/3/015
6. W. Shockley, *Electrons and Holes in Semiconductors* (Robert E. Krieger Publishing Co., New York, 1976)
7. S.M. Sze, K.K. Ng, *Physics of Semiconductor Devices*, 3rd edn. (Wiley, 2007)
8. P.J. Collings, Simple measurement of the band gap in silicon and germanium. Amer. J. Phys. **48**, 197–199 (1980)
9. J.W. Precker, M.A da Silva, Amer. J. Phys. **70**(11) (2002)
10. C.W. Fischer, Elementary technique to measure the energy band gap and diffusion potential of pn junctions. Amer. J. Phys. **50**, 1103–1105 (1982)
11. W. Bludau, A. Onton, W. Heinke, Temperature dependence of the band gap of silicon. J. Appl. Phys. **45**, AIP, 1846–1848 (1974)
12. R.A. Abram, G.N. Childs, P.A. Saunderson, Band gap narrowing due to many-body effects in silicon and gallium arsenide. J. Appl. Phys. **45**, 1846–1848 (1974)
13. Y. Pan, M. Kleefstra, Minority carrier transport equations in heavily doped silicon including band tail effects at thermal equilibrium. JPC **17**, 6105–6125 (1984)
14. P. Horowitz, W. Hill, *The Art of Electronics*, 2nd edn. (Cambridge University Press, Cambridge, 1989)

Chapter 2
Review of Metal-Semiconductor Junctions

Abstract This chapter reviews the important founding principles of MSJ diodes. It sets the stage for the presentation of parameter extraction methods in the following chapters. The focus is primarily on MSJ diodes since all solid-state devices involve such junctions. In the interest of presenting a more consistent discussion, the scope is focused on devices constructed on p-type silicon.

2.1 A Historical Perspective

The history of the semiconductor diode is a fascinating journey that began with the accidental discovery of peculiar, asymmetrical conduction in solid-state materials in the late 19th century. Ohm's Law had become a *de facto* expectation for conductors in the solid state. It was, therefore, puzzling when certain solid materials exhibited deviations from this trend under certain conditions. It was known that external factors, notably incident illumination and temperature changes, could heavily influence the current flowing through these materials at a given applied potential. These odd materials often formed unwittingly on only specific crystalline substrates.

The concept of bias, like Ohm's Law, was also well-established due to developments in thermionic emission or vacuum tube devices. The term 'bias' implies arranging the conditions of a circuit externally in a way that encourages a desired operation. For these solid-state devices or diodes, the term "forward bias" was adopted to mean the mode of connection that resulted in higher conductivity, whereas "reverse bias" indicated lower, often undetectable, conductivity. Thus, the new devices demonstrated the phenomenon of rectification. Early diodes swiftly found application as replacements for metal-grain coherers as radio wave detectors with higher efficiencies and lower detection thresholds. It was soon realized that they could be adopted in other fields and this saw a rise in their popularity.

The application of a semiconductor diode in a circuit requires metal terminals attached to some points or regions of the solid materials to interact with it through voltage and current. These simple descriptions of the semiconductor devices, pinpoint the peculiarities of solid-state diodes that remain the subject of intense study to this day. These are summarized below.

1. The diode material type, broadly referred to as "semiconductors", is crucial. Initially, diode behaviors were observed in elemental materials like Se and Ge, as well as compound materials such as CuO and PbS, among others.
2. The terminal metal type plays a significant role. Early devices exhibited asymmetrical conduction exclusively when specific metals were in contact with a particular semiconductor.
3. Despite the aforementioned peculiarities, it was observed that achieving optimal rectification behavior required some degree of adjustment even with the appropriate metal-semiconductor combination. Hence, the nature of the contact is crucial.
4. Later on, it became evident that the behavior of semiconductor junctions was also crucial. It was discovered that the conductivities of certain materials couldn't be explained solely by the presence of electrons but rather by their deficiency. This led to the proposal of "n-type" and "p-type" materials, for which further observations showed that the junctions between them displayed asymmetrical conduction. It was important to ensure that the metallic lead-outs were carefully chosen to prevent rectification at those points.

Today, with hindsight, we know that these early observations describe what is known as metal-semiconductor or Schottky junctions [1], rather than their close relative, the p-n junction diode that came much later. Nonetheless, there remain considerable challenges to completely understand their operation.

This book focuses primarily on the above aspects of solid-state diodes, the important characteristics and the laws describing them, and the most current ways of extracting their parameters. Our interest is not to reproduce much of what is already well-known and well-described in other literature but to dedicate ourselves largely to new techniques for characterizing semiconductor junction devices, particularly Schottky junctions. This remains an important endeavor since all solid-state devices rely on metal-semiconductor interfaces.

The metal-semiconductor (MS) or Schottky diode forms when specific metals create two junction interfaces with a semiconductor: one rectifying and the other serving as an Ohmic, ideally zero-resistance electrical lead-out. Since most solid-state devices rely on metal-semiconductor interfaces, they can potentially be Schottky junctions, leading to extensive literature and ongoing research in new semiconductors and interface metallization [2–5]. Schottky diodes find applications in various fields, including fast-recovery rectifiers, RF, thermal and optoelectronic detectors, transistors, and more.

2.2 The Schottky Barrier

The earliest systematic description of asymmetrical conduction in a solid material was in a device in which the electrical contact occurred at a single "sweet" point between a metal and the semiconductor. It was realized the surface conditions existing at the contact point determined the behavior of the diode. Locating this point required trial and error, hence the variability and low repeatability in the characteristics of these devices.

The first real attempt to model the operation of the semiconductor diode was established in the 1930s using the band theory of solids [6, 7]. This opened the gates to the theoretical exploration of these devices. Quantum mechanics, which developed around the same time, provided a powerful new tool that advanced the study of these solid-state diodes by explaining phenomena such as rectification, albeit difficulties remained, such as explanations of the direction of the observed current. In 1938 Schottky and Mott independently postulated the existence of a potential energy barrier that influenced drifting and diffusing electrons.

Drift refers to the movement of electrons (charge carriers in general) under an applied bias potential. Diffusion, on the other hand, refers to their movement down a concentration gradient. Mott theorized that the internal potential barrier resulted from the work function difference between the semiconductor and the contacting metal. In his model, the charges exist at the boundaries of the potential barrier such that the electric field is constant within the barrier. On the other hand, Schottky presumed a constant charge density in the potential barrier region, which implied a linearly increasing electric field from one side of the barrier to the other. It was not until 1942 that Bethe developed TE theory, which satisfactorily explained the rectification currents using the idea of electrons injected (emitted) into the metal through a temperature-critical (i.e., "thermionic") process [8]. Further work with point contacts culminated in the development of the point-contact transistor which set off the solid-state electronics revolution.

TE theory, which prevails to this day, makes three key assumptions. Firstly, the energy of the potential barrier, characterized as the barrier height ϕ (in eV) or $q\phi$ (joules), is significantly larger than the thermal energy kT. Thus, the current density is calculated only for electrons with sufficient energy to overcome the potential barrier. Secondly, thermal equilibrium is established at the emission plane and, lastly, the net current flow does not affect this equilibrium.

2.2.1 MS Surfaces

Point-contact junctions suffer from issues of repeatability, reliable fabrication, and predictable operation. In the 1940s devices using semiconductor surfaces interfaced with evaporated metallic films arose that were easier to make and more robust in their characteristics. However, the challenges posed by surface conditions on the MS interfaces remained.

Both metals and semiconductors are crystalline in the bulk but their surfaces are thought of as defects since the termination of the long-range ordering occurs there [9]. The interrupted arrangements of the atoms at the surface are highly dependent on the cleaving cross-section, which directly impacts the surface energy distribution, a phenomenon known as relaxation. For instance, the low-energy (111) surfaces in silicon and germanium have a high DOS. The effects of relaxation extend several lattice parameters below the surface into the bulk and play an important part in the operation of MS devices. Also, metal and semiconductor surfaces are no longer pristine within fractions of a second after being exposed to ambient conditions. This is the result of prevailing environmental factors such as oxidation and other contamination. Detailed discussions on the effects of the disruption of crystalline ordering at the surface can be found in many literature sources [10].

One of the more interesting effects is that the energy states can either be discrete or a continuum within the band gap of the material. These states cannot be treated like conduction or valence states. In silicon, the calculated DOS is shown in Fig. 2.1.

The distribution of surface states in p-Si is shown in Fig. 2.2. The diagram assumes that the FD distribution is a step function, s.t below E_F all states are filled, and unfilled above it. The electron affinity χ is the energy needed to remove an electron from the conduction band to the vacuum level. At E_F, there is a 50% probability of finding an electron. In metals, the Fermi level lies within the conduction band, which means that there are free electrons available for conduction. In contrast, semiconductors have energy bands that include the valence band (where electrons are bound) and the con-

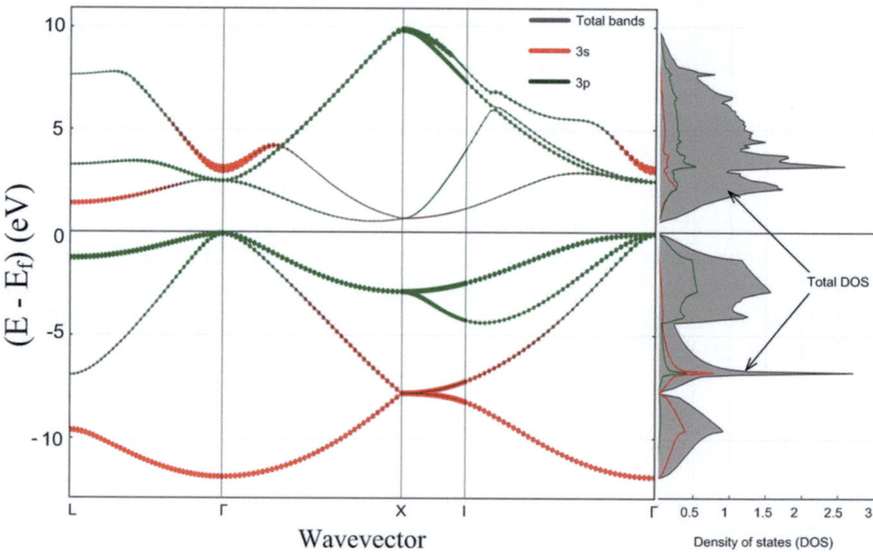

Fig. 2.1 The calculated density of states of silicon formed by the overlap of the 3s^2 and 3p^2 electrons. This DOS distribution is confirmed in several sources using XPS measurements. The lowest bandgap is 1.12 eV

Fig. 2.2 Depiction of the distribution of surface states in p-Si in the low-energy (111) cleaved surface, assuming a step FD distribution. The work function, φ, is the energy needed to raise an electron from E_F to the vacuum level. It has both a surface aspect due to surface dipoles, and a volume aspect due to the periodic potential and electron interactions

Fig. 2.3 Depiction of the energy levels in an idealized metal. For a metal, the electron affinity, χ, is the same as φ

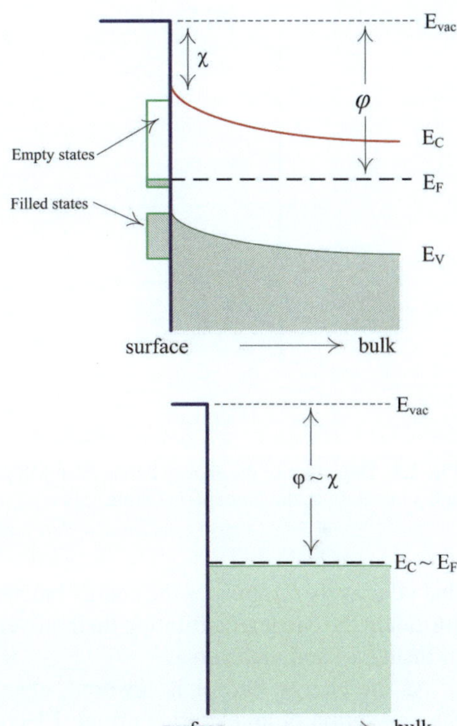

duction band (where electrons can move freely). Figure 2.3 shows the energy band diagram of an idealized metal with a pristine surface. Figure 2.4 illustrates the importance of the relative sizes of the work functions of the metal and the semiconductor in determining the electrical behavior of the junction.

2.2.2   Conduction Mechanisms

Simply put, an MSJ is formed when a metal and a semiconductor come into contact. However, the reality is more complex, for some of the reasons in Sect. 2.1. The behavior of this junction depends on the relative positions of E_F and the energy bands of the semiconductor. In the case where the metal Fermi level falls within the bandgap of the semiconductor, a Schottky barrier is formed. The metal and semiconductor energy bands align, and charge transfer occurs at the interface. The interface significantly influences the electronic behavior of the device. The Schottky emission mechanism involves the flow of electric current across the MS interface when subjected to an electric field. When an external electric field is applied to the metal-semiconductor or metal-insulator junction, it generates an electrostatic force

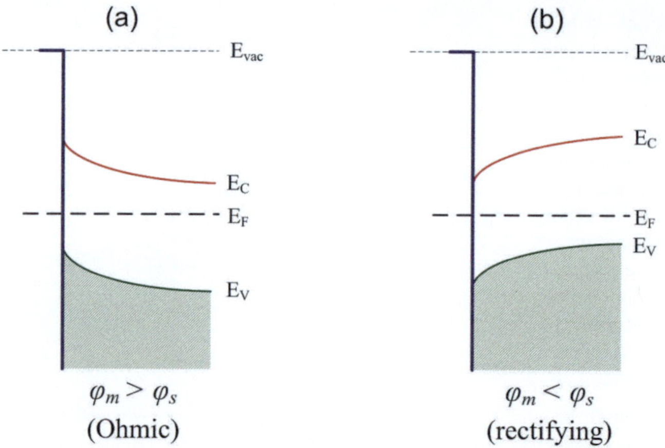

Fig. 2.4 Depiction of the energy bands in an MS junction formed between a p-type semiconductor and a metal. In **a**, the junction is Ohmic since $\varphi_m > \varphi_s$. In **b**, a rectifying junction

that effectively diminishes the energy barrier. Consequently, electrons find it easier to surmount the barrier, facilitating their movement from the metal to the semiconductor or insulator, and vice versa.

As the energy barrier is lowered, electrons traverse the interface, resulting in the generation of an electric current. Electrons can tunnel through the barrier from the metal side to the semiconductor/insulator side, and vice versa. There is another conduction mechanism the Poole-Frenkel effect [11], which though similar to the Schottky effect, requires less thermal energy to transition into the conduction band. This reduction in energy demand is attributed to the electron being drawn by the electric field itself, effectively contributing a portion of the required energy. Consequently, the electron necessitates a smaller thermal fluctuation to make this transition, leading to heightened mobility and increased frequency of movement.

The Schottky effect pertains to the reduction of the energy barrier in contact-limited conduction scenarios between a metal and an insulator due to electrostatic interactions arising at their interface in the presence of an electric field, whereas the Poole-Frenkel effect dominates where there is bulk-limited conduction, where the predominant conduction process unfolds within the material's interior. The foregoing discussion assumes that another very important phenomenon, referred to as *impact ionization*, is absent [12].

The Schottky barrier acts as a barrier to electron flow from the metal to the semiconductor s.t. only electrons with sufficient energy, e.g., from thermal excitation, can cross from the metal to the semiconductor. The height of the energy barrier, Φ, is the difference between the metal Fermi level, $E_{F,m}$ and the semiconductor valence band edge E_V in a p-type semiconductor. Figure 2.4 shows the conditions under which p-type/metal junctions are either rectifying or Ohmic.

In the interest of presenting the content more concisely, our focus going forward is on junctions of p-type semiconductors with metals with work functions φ_m smaller than that of the semiconductor, φ_s.

2.2.3 MS Interface Energy States

As mentioned above, the junction between a metal and a semiconductor has imperfections that modify the band diagrams shown above. In reality, the barrier height is less dependent on the type of metal forming a junction with a given semiconductor, and more on the presence of any interfacial, and usually very thin, insulating layer. This layer adds surface energy states which significantly modify the electrical characteristics of the junction. The conduction of electrons across the junction in forward bias occurs through the phenomenon of quantum mechanical tunneling through the thin interfacial layer. The commonly encountered energy band diagram is shown in Fig. 2.5, under the condition of applied forward bias, V.

2.2.4 Oxide Inter-Layers and Surface States

In Fig. 2.2 it is apparent that the barrier height at the junction in the absence of an applied bias is given by

$$\Phi = \varphi_m - \chi_s.$$

(2.1)

However, in reality, the junction is imperfect and usually has a thin insulating layer of oxide material formed from one or both of the surfaces making up the junction. The assumption is that the junction is so thin that the electrons at the junction can tunnel through the energy barrier presented by the oxide layer.

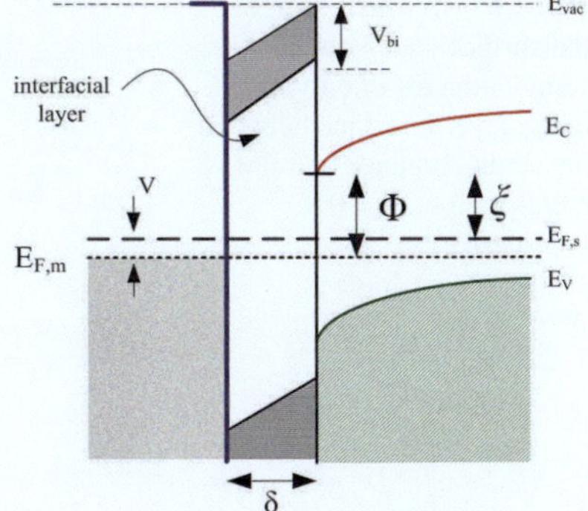

Fig. 2.5 The energy band diagram of an MSJ with a thin interfacial layer of thickness, δ. Quantum mechanical tunneling occurs at the interface. The energy difference between the bottom of the conduction band and the Fermi level is ξ. The flat-band barrier height can be taken as Φ_0

Equation 2.1 suggests that the barrier height relies on the work function of the metal, i.e., the type of metal being used, but practical experiments have demonstrated in numerous instances that φ_m is not the sole and most significant determinant of Φ. Fermi level pinning describes a scenario in which the band bending at the interface of a semiconductor and a metal remains relatively unaffected by substantial variations in the metal's work function, leading to a consistent behavior despite significant differences in metal properties [13, 14]. This discrepancy can be attributed to the presence of what is commonly referred to as surface, or interface, states that are located on the surface of the free semiconductor. These energy states alter the charge neutrality condition at the junction in a manner that ensures an electrically neutral junction by accounting for the net charge contributed by the metal, semiconductor, and interface state charges.

In the absence of surface states, the charge neutrality condition is defined by the equilibrium between the charge Q_m on the metal and a corresponding but opposite dipole charge Q_d originating from uncompensated donors. When surface states are present the charge neutrality condition undergoes modification to incorporate an additional term, denoted as Q_{ss}. Figure 2.6 illustrates the distribution of the surface states at the surface of the semiconductor. The neutral energy level, E_{ss}, of the surface states, is close to E_F at thermal equilibrium and is typically assumed to be the same as E_F in the absolute-zero "step" approximation in the Fermi-Dirac distribution [9]. Thus, the surface states below E_F are assumed to be filled, and those above E_F are assumed to be empty. It is customary to measure the barrier height according to

$$\Phi \approx E_g - E_{ss}. \tag{2.2}$$

When the density of surface states is very large i.e., approaching $10^{17}\,\text{eV/m}^2$, then $E_{ss} \approx E_F$, effectively pinning the Fermi level. Alternatively, for a p-type MS junction with applied bias V, the energy level of the interface states is given by [15, 16]

$$(E_{ss} - E_V) = \Phi - V \tag{2.3}$$

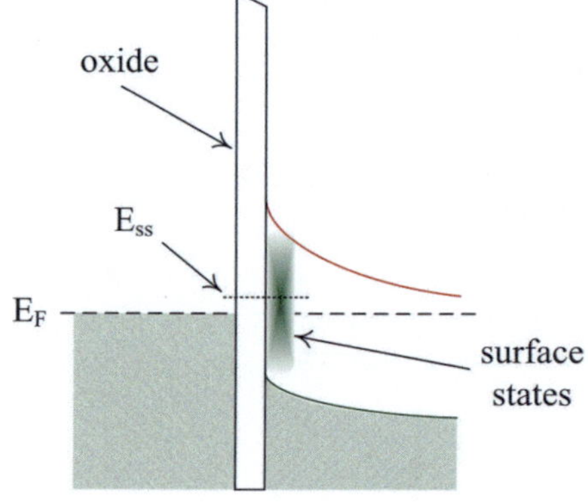

Fig. 2.6 An illustration of the surface states on the semiconductor due to an oxide layer interface between the semiconductor and the metal. In the $T = 0\,\text{K}$ approximation, $E_{ss} \approx E_F$

It is important to understand that the distribution of the surface states is rarely uniform, giving several charge recombination mechanisms at the interface [17, 18]. The mechanisms produce certain well-documented effects, such as during photo-conduction, that are discussed in a later chapter. In summary, the neutrality conditions can be expressed as follows:

$$Q_m + Q_d = 0, \quad \text{no surface states, and}$$
$$Q_m + Q_d + Q_{ss} = 0, \quad \text{with surface states.} \tag{2.4}$$

In practice, the net charge contribution of the surface states depends on their precise location relative to the Fermi level in the energy bands. If they are above, then their net polarity is positive. This leads to a smaller Q_d, a narrower depletion width w, and, consequently, less band bending and a reduced barrier height.

2.3 Field and Bias Dependence of Barrier Height

As mentioned above, the presence of an oxide interfacial layer influences the electric field within the oxide and modifies the barrier height. If one assumes a continuous distribution of interface states such that the density of interface states is constant, then the barrier height reduction is directly proportional to the electric field due to an externally applied bias in the semiconductor [10] i.e.:

$$\Phi = -\left(\frac{\delta \varepsilon_s}{\varepsilon_i + q\delta D_{it}}\right)\mathcal{E}_m + \Phi_0 = -\alpha \mathcal{E}_m + \Phi_0, \tag{2.5}$$

where α is the term in brackets, the dielectric permittivities of the semiconductor and the interfacial layer are ε_s and ε_i, respectively; q is the electronic charge, D_{it} is the density of the interface states and $\mathcal{E}_m$ is the maximum electric field experienced at the top of the energy barrier. As long as D_{it} remains constant, Eq. 2.5 is valid for any applied bias and means that the barrier height is reduced from its zero-bias value in a way that depends on the maximum electric field. A more useful form of Eq. 2.5 arises from the knowledge of the net potential drop V_d across the depletion region i.e.:

$$V_d = \frac{1}{2}\frac{\varepsilon_s \mathcal{E}_m^2}{q N_d}, \tag{2.6}$$

where N_d is the active donor concentration in the region. This potential difference, or diffusion potential, drives the diffusion across the depletion region. Referring to Fig. 2.5, one sees that in the presence of an applied potential, the diffusion potential can be written as

$$V_d = \Phi_0 - V - \xi, \tag{2.7}$$

where ξ is the energy difference between the bottom of the conduction band and the Fermi level, and Φ_0 is the "flat-band" barrier height. Consequently, Eq. 2.5 can be written in terms of the applied potential. However, it is usual to take into account the effects of the region in which the electron density is negligible compared to the donor density, N_d, by modifying the applied bias by (kT/q), such that

$$V_d - \frac{kT}{q} = \frac{1}{2}\frac{\varepsilon_s \mathcal{E}_m^2}{q N_d}. \tag{2.8}$$

Combining Eqs. 2.5, 2.7, and 2.8, allows the net barrier height under external bias to be written more generally as

$$\Phi = \Phi_0 - \left[\left(\frac{2q\alpha^2 N_d}{\varepsilon_s}\right)\left(\Phi_0 - V - \xi - \frac{kT}{q}\right)\right]^{1/2}. \tag{2.9}$$

One could write Eq. (2.9) as

$$\Phi = \Phi_0 - \beta(\gamma - V)^{1/2} \tag{2.10}$$

by setting $\gamma = (\Phi_0 - \xi - kT/q)$ and $\beta = (2q\alpha^2 N_d/\varepsilon_s)^{1/2}$, which are essentially constants. Then, expanding the term $(\gamma - V)^{1/2}$ using the binomial theorem:

$$\begin{aligned}(\gamma - V)^{1/2} &= \gamma^{1/2} - \frac{1}{2}\gamma^{-1/2}V - \frac{1}{8}\gamma^{-3/2}V^2 - \frac{1}{16}\gamma^{-5/2}V^3 + \cdots \\ &= \gamma^{1/2} - \frac{1}{2}\gamma^{-1/2}V + O(3). \end{aligned} \tag{2.11}$$

It can be argued that the terms containing powers of V greater than three in this result become increasingly smaller and can be ignored since $\gamma < 1$. Additionally, V is generally measured in the low-forward-bias region, further justifying the approximation. Substituting Eq. 2.11 into Eq. 2.10 then gives an expression that suggests that the variation of barrier potential with applied bias is approximately linear i.e.:

$$\Phi \simeq a \pm bV, \tag{2.12}$$

where $a = (\Phi_0 - \beta\gamma^{1/2})$ and $b = (\beta\gamma^{-1/2}/2)$ are also essentially constants. The sign of V is negative for forward bias and positive for reverse bias. As will be seen in a later section, the variation of the barrier height of Schottky diodes with applied bias is approximately linear. It is worth noting that the term γ as defined conveys the temperature term and, therefore, can give insights into another important parametric variation, the temperature dependence of the barrier height. Loosely speaking, one could explore this as follows.

2.4 Image-Force Barrier Lowering

When a charge is located sufficiently near the surface of a metal, it induces an image charge of the opposite sign inside the metal. As a result, there is an attractive Coulomb force, $\bar{F} = \bar{\nabla} V(\bar{r})$, between the charge and its image, where V is the potential energy of the two-charge system.

An electron above the metal thus experiences a force that is directed toward the metal and assists the injection of the electron into the metal. The net result is a barrier potential energy barrier from the electron's perspective. The amount by which the barrier is lowered is equal to the change in potential energy. This is known as *image-force barrier height lowering* (IFBL).

In a practical device, the charge resides in the semiconductor in contact with the metal. Assuming that a perfect contact exists at the MS junction, the interaction of the two charges occurs at twice the distance of the charge from the metal surface. A further simplification is possible by noting that the interaction occurs perpendicularly to the junction, say in the direction x so that $\bar{r} = x$, and $F(= dV/dx)$. Consequently,

$$V(x) = \int_x^\infty \frac{dx}{4\pi\varepsilon_0\varepsilon_s(2x)^2}dx = \frac{q}{16\pi\varepsilon_s'x} \tag{2.13}$$

or

$$V(x) = x\mathcal{E}(x), \tag{2.14}$$

where $\varepsilon_s' = \varepsilon_0\varepsilon_s$ is the effective dielectric permittivity of the semiconductor, and ε_0 is the dielectric permittivity of free space. Typically, ε_s' is stated as a high-frequency parameter owing to the high electron mobility in the semiconductor. This assumes that the semiconductor does not have sufficient time to be ionized, otherwise, the static permittivity would be used.

Numerous experiments have shown that for p-Si, the static and high-frequency dielectric permittivities are, for practical purposes, barely indistinguishable at $11.7\varepsilon_0$ and $(12.0 \pm 0.5)\varepsilon_0$, respectively. The amount of barrier height lowering $\Delta\Phi$ due to the image forces can be calculated using Eqs. 2.8 and 2.14. If the charge is to be positioned in a stable and immovable point above the metal the resultant electric field is zero. If this occurs at a depth x_m into the metal, then

$$\Delta\Phi = x_m\mathcal{E}_m + V(x_m). \tag{2.15}$$

IFBL has become a particularly significant consideration in light of the new and growing field of 2D materials for electronic devices. It is now accepted that forming low-resistance contacts in these materials requires an accurate expression of image force barrier-lowering, with many different approaches currently being employed to study it. Some solve Poisson's equation in a variety of surface orientations of the metal-semiconductor interface [19–21]. These recent investigations show that the dielectric permittivity of the surrounding oxide plays a significant role in the image forces, and thus in the barrier height.

2.5   Series Resistance in MS Devices

A peculiar if problematic parameter in the characteristics of a junction device is the series resistance, R_s. Arguably, the series resistance is almost singularly responsible for the deviations observed from non-ideal behavior in the forward characteristics of the MS diode. An insulating interfacial layer contributes to R_s and may arise from an inevitable native oxide layer or be introduced unwittingly by the fabrication process. It is then more appropriate to refer to the device as a metal-insulator-semiconductor (MIS) diode [22, 23].

Apart from the effects of the natural resistivity of the semiconductor, the device geometry, and the nature of the contact metallization, R_s may also emerge during the formation of the MS junction due to differences in electronic properties that introduce an energy barrier that hinders charge transfer. At the metal and semiconductor interface, surface states, oxides, and contamination can establish barriers that hinder the movement of charge carriers. These surface effects directly contribute to the overall series resistance by obstructing the smooth flow of current by acting as traps.

Phonon scattering events due to impurities and lattice defects also manifest series resistance since the overall effect is reduced charge carrier mobility. R_s, in combination with frequency-selective measurement methods, can lead to useful estimates of the density of the interface states. In any event, the importance of R_s is such that for many measurements to be meaningful, they must be R_s-compensated [24].

2.6   Interfacial Capacitance

The total capacitance at the junction in a real device is a combination of diffusion capacitance, C_a, and the depletion layer capacitance, C_b.

Diffusion capacitance dominates in forward bias, V i.e.:

$$C_a = \frac{q^2 n_i^2 L_n}{N_a} \left(\frac{e^{qV/\alpha kT}}{kT} \right), \tag{2.16}$$

where L_n is the diffusion length, α is a negative dimensionless constant, and N_a and n_i are, respectively, the non-ionized acceptor and intrinsic carrier concentrations [25]. Most MS diodes are operated in the low-forward bias region, where V is small. The diffusion capacitance varies approximately linearly with the applied bias under this condition. Equation 2.16 can be written in an alternative activation energy form by setting $E_a = (qV/\alpha)$ and $A^* = q^2 n_i^2 L_n / N_a$ i.e.:

$$C_a = A^* e^{E_a/kT}. \tag{2.17}$$

The differential depletion-layer capacitance, $C(= dQ/dV)$, can be estimated under reverse bias V_r inside the semiconductor's depletion or space-charge region using the diffusion potential (Eq. 2.8) [10, 26]. For a p-type MS junction with a cross-sectional area A,

$$C_b = A \left(\frac{\varepsilon'_s q N_a}{2(\Phi_0 - \xi + V_r - kT/q)} \right)^{1/2}, \tag{2.18}$$

where N_a is the density of acceptors, which are assumed to be non-ionized. Practical measurements of capacitance over reverse bias allow Eq. 2.18 to be rearranged for a linear plot of C^{-2} versus V_r i.e.:

$$C_b^{-2} = \mu_A V_r + \mu_B, \tag{2.19}$$

where the constants μ_A and μ_B denote the gradient and intercept, respectively, and are given by

$$\mu_A = \frac{2}{\varepsilon'_s q A^2 N_a}, \quad \text{and} \quad \mu_B = \frac{2}{\varepsilon'_s q A^2 N_a} \left(\Phi_0 - \xi + V_r - \frac{kT}{q} \right).$$

In reverse bias measurements, diffusion capacitance is negligible in comparison to depletion capacitance and can be ignored. Therefore, for a given device of known cross-sectional area, the non-ionized acceptor density, N_a, can be estimated from the gradient, and the zero-bias barrier height Φ_0 can be estimated at the measurement temperature T from the intercept, i.e.:

$$\Phi_0 = \xi + \frac{kT}{q} + \frac{\mu_B}{\mu_A}. \tag{2.20}$$

Figure 2.7 shows the typical plot of Eqs. 2.18 and 2.19.

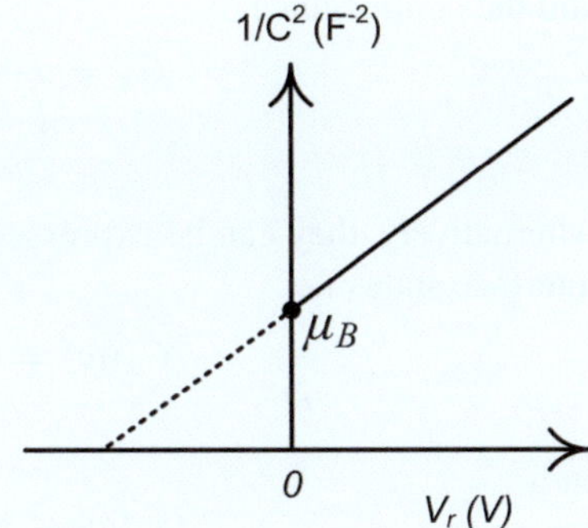

Fig. 2.7 A representation of the typical plot of C^{-2} versus the applied reverse bias for n-type Schottky diodes. The plot can be drawn reversed in p-type diodes. The non-ionized acceptor density N_a (or donor density N_d, for n-type material) is estimated from the slope, whereupon the barrier height follows from the intercept, μ_B

2.6.1 The Impact of Series Resistance

Reverse-bias capacitance-voltage measurements of barrier height and other related parameters will generally disagree with the results of other methods due to the effects of R_s. Capacitance is a reactive parameter that is measured using a frequency-voltage setup. However, the presence of resistance in any circuit configuration with the capacitor, as simulated by realistic MS junctions, presents phase shifts at different frequencies that must be considered.

The measured capacitance is further complicated by interface and impurity states that trap charge carriers dynamically in the frequency domain. Therefore, to be meaningful, measured capacitance and other reactive parameters like conductance, G, must be compensated for R_s. Therefore, it is more meaningful to write the true, measured junction capacitance as $C_a(\omega)$, to denote the frequency dependence and adjustment for the effects of R_s.

Figure 2.9 shows the effect of series resistance on the measured capacitance of a typical MS diode. It is clear that R_s has a significant effect on the capacitance. Figure 2.10 shows the effect of R_s on the reactive conductance of a typical MS diode. The compensated conductance exhibits peaking, which highlights the presence of interface states in the junction.

Figure 2.8 shows the equivalent circuit of the total depletion capacitance across the MS junction in reverse bias [27, 28]. From the simplified equivalent circuit shown, The adjusted capacitance C_a, and conductance G_a at the junction at a measurement angular frequency ω are functions of the interface state distribution. In terms of the measurable parallel capacitance:

$$C_a = \frac{C_m\left[1 - \frac{C_m}{C_{ox}}\right] - \frac{1}{C_{ox}}\left(\frac{G_m}{\omega}\right)^2}{\left(\frac{G_m}{\omega C_{ox}}\right)^2 + \left[1 - \left(\frac{C_m}{C_{ox}}\right)\right]^2},$$
(2.21)

and the conductance

$$G_a = \frac{G_m}{\left(\frac{G_m}{\omega C_{ox}}\right)^2 + \left[1 - \left(\frac{C_m}{C_{ox}}\right)\right]^2}.$$
(2.22)

Alternatively, they can be expressed in terms of τ, the trapping time constant of the interface states i.e.:

$$C_a(\omega) = C_d + \frac{C_{it}}{2}\left[\frac{\tan^{-1}\omega\tau}{\omega\tau}\right],$$
(2.23)

and

$$\frac{G_a(\omega)}{\omega} = \frac{qD_{it}A}{2}\left[\frac{\ln(1 + \omega^2\tau^2)}{\omega\tau}\right].$$
(2.24)

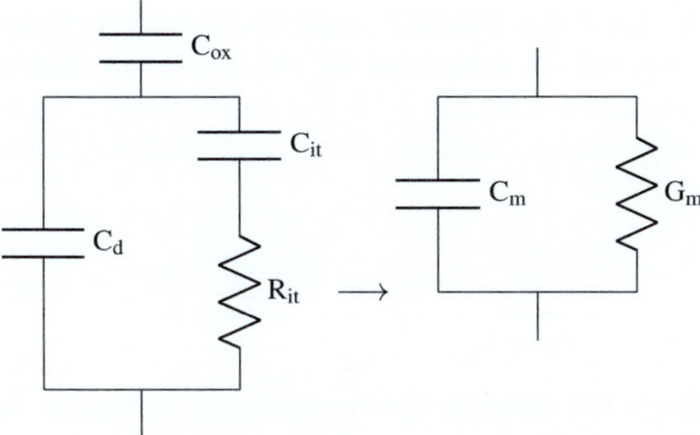

Fig. 2.8 The equivalent circuit (left) of the MS junction in reverse bias showing the contribution of depletion capacitance, C_d the oxide capacitance, C_{ox}, and the interface states capacitance and resistance, C_{it} and R_{it}, respectively. The simplified equivalent (right) shows the overall measured capacitance C_m and conductance, G_m

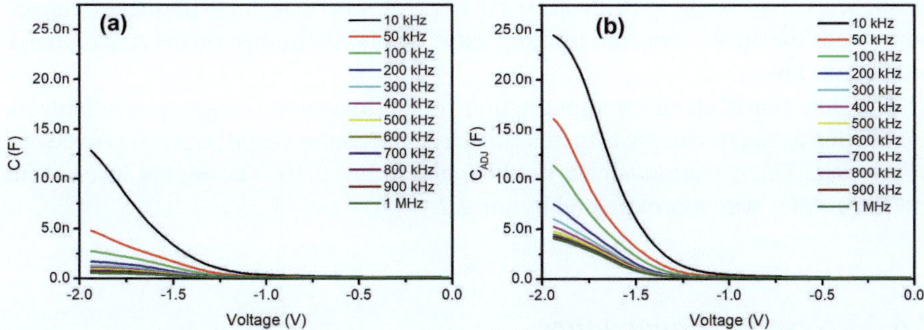

Fig. 2.9 Graphical illustration of the importance of R_s compensation of the measured depletion capacitance in a real MS junction diode. In **a**, a plot of the measured capacitance data directly from the instrument. In **b**, the same data is plotted after R_s compensation

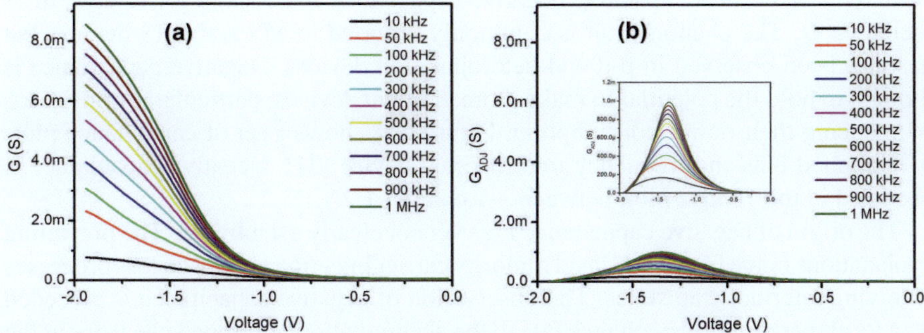

Fig. 2.10 A similar graphical illustration of the need for R_s compensation of the reactive conductance of a real MS junction diode. In **a**, a plot of the uncompensated, and **b** compensated conductance. The magnified inset figure in (**b**) emphasizes peaks that indicate the strong frequency dependence of interface states

In Sect. 2.3, it was stated that interface states at the junction follow a distribution. Therefore, it is more usual to determine the worst-case interface density by taking the maximum conductance using Eq. 2.24. The utility of Fig. 2.10 is thus to illustrate peaks (pk) in the adjusted conductance. For a given peak (i.e., for a specific ω), this can easily be shown to be

$$\underbrace{D_{it}}_{\text{worst-case}} = N_{ss} = \frac{(G_a(\omega)/\omega)_{pk}}{0.402\, q\, A}. \tag{2.25}$$

Interface states are felt through their reactions to alternating current signals as sluggish traps that curtail charge mobility. Admittance spectroscopy offers a means to gauge their behavior across frequencies, yielding crucial insights into their distribution. As the stimulation frequency increases at a constant amplitude, the ability of interface states to trap charge diminishes, causing a decrease in junction capacitance [29]. In short, the interface state trapping activity cannot keep up with high-frequencies. This scenario is illustrated in Fig. 2.9, where the high-frequency capacitance tends towards a constant, limiting value that is determined by the space-charge (depletion) layer.

The reduction of charge-trapping action implies that more charges are available for conduction as the frequency is increased. Thus, the conductance ($G_a(\omega)$) is expected to increase. This situation is clearly illustrated in Fig. 2.10. The energy level of the interface states was approximated by Eq. 2.2.

2.6.2 Negative Capacitance

In recent years, a growing body of evidence suggests that, under forward bias and low frequencies, a wide range of devices exhibit negative capacitance [30]. By definition, since $Q = CV$ for a capacitor, negative capacitance means that a decrease in V increases Q. The phenomenon is commonly reported in MS and MIS devices but has also been observed in p-n and heterojunction devices. Negative capacitance is thought to hold the potential to make more efficient devices, particularly transistors, by reducing their power consumption. Figure 2.11 shows a set of capacitance plots over applied bias and frequency in a reported device [31]. Negative capacitance is observed in the 10 kHz plot, between -1.6 and -1.2 V.

The origin of negative capacitance is as yet not clearly established. The prevailing explanations range from high-level minority carrier injection to impact-loss processes involving interface trap states. The observation of negative capacitance is preceded by a local increase in capacitance due to the accumulation of trapped electrons at the interface trap sites [32]. It is thought that negative capacitance is favored by metal-oxide contacts with high work-function metals where migrating oxygen vacancies accumulate and pin the Fermi level.

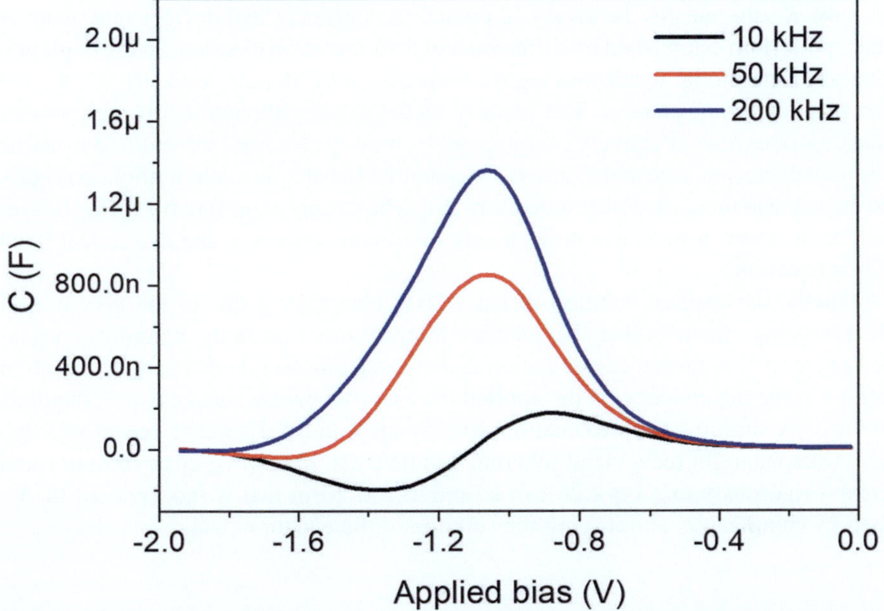

Fig. 2.11 A graph showing the bias and frequency dependence of negative capacitance in a device made from graphene oxide-doped Coumarin p-Si heterojunction diode [31]

2.7 The Thermionic Emission Model

In the TE model, the Schottky diode current, denoted by I, exhibits a cyclical[1] dependence on various parameters such as the applied voltage V, absolute diode temperature T, effective barrier height ϕ, and the ideality factor n [10, 22, 33]. Additionally, in the presence of appreciable diode series resistance R_s, the relationship becomes more intricate. Mathematically, the diode current in the non-ideal (i.e., non-zero R_s) TE model can be described by the following expression:

$$I = I_o \exp\left(\frac{qV - IR_s}{nkT}\right)\left\{1 - \exp\left(-\frac{qV - IR_s}{kT}\right)\right\}, \tag{2.26}$$

where

$$I_o = AA^*T^2 \exp\left(-\frac{q\phi}{kT}\right), \tag{2.27}$$

and where q represents the electronic charge, k is the Boltzmann constant, A represents the diode area, and A^* is the Richardson constant. The term I_o is the reverse saturation current of the device.

---

[1] The term "cyclical" here meaning that I depends on itself in combination with another parameter.

Like R_s, the ideality factor n is a pivotal parameter in real devices that nuances the equilibrium established by diffusion and drift forces on electrons in the depletion region of the diode. It depends on the semiconductor doping level, its purity, and the manufacturing process. The ideality factor may be thought of as a parameter that describes how effectively charge carriers, mainly electrons, are emitted from the semiconductor into the metal across the potential barrier. As such, it might correctly be expected to increase with temperature and to be strongly correlated with the barrier height. In short, n indicates how closely the diode adheres to the theoretical ideal diode equation.

Finally, the explicit solution of Eq. 2.26 is challenging due to the presence of the foregoing effects besides the junction temperature, especially R_s and its circular reference to I. A further complication is that these inherent device parameters often exhibit some dependency on the applied bias. Consequently, most existing methods use heavily simplified approximations to estimate Φ and n. However, recently Ocaya and Yakuphanoğlu recognized inherent symmetry in the device characteristics and proved rigorously that Eq. 2.26 can be reduced to form that is independent of R_s, thereby significantly simplifying the solution of the equation [34].

2.7.1 Richardson's Constant

In the contemporary TE equation given by Eq. 2.26, the term A^* is known as the modified Richardson's constant, after physicist Owen W. Richardson who first proposed the approach [35]. In the original work, A was derived by using energy minima in materials that express a work function, φ. The current density J in such materials at a temperature T can be written in the activation energy form:

$$J = AT^2 e^{-\varphi/kT}, \tag{2.28}$$

or

$$\ln\left(\frac{J}{T^2}\right) = -\left(\frac{\varphi}{k}\right)\frac{1}{T} + \ln A. \tag{2.29}$$

Richardson's plot results when $\ln(J/T^2)$ is plotted against $1/T$. It is expected to be a straight line with a negative gradient and an intercept that leads to A.

The constant A characterizes the thermionic emission process in the materials where inter-level emissions of fermions occur. This is loosely applicable to metals and semiconductors. The original work presumed lossless thermionic emissions, which in solid-state materials imply perfect contact at the interfaces. Such a contact might implement a Schottky junction with conditions that are not fully achievable, thereby limiting the utility of Eq. 2.29. In Schottky barrier contacts, the imperfections of the interfaces lead to several spatial effects, collectively referred to as 'barrier height inhomogeneities' (BHI).

2.7.2 *Modified Richardson's Plot*

The reverse saturation current in Eq. 2.27 has the form of Richardson's formula and enables the estimation of A^*, assuming that ϕ is constant. In reality, deviations from the TE model are observed as a consequence of BHI.

The usual corrective treatment is to assume that the measured barrier heights ϕ are normally distributed around a mean value barrier height $\bar{\phi}$. A Gaussian function, $P(\phi)$ can thus be defined [36]:

$$P(\phi) = \frac{1}{\delta\sqrt{2\pi}} e^{-(\phi-\bar{\phi})^2/(2\delta^2)}, \tag{2.30}$$

where δ is the standard deviation. This approach applies to all the measured variables in the Schottky device IV and not just ϕ. For instance, a series of IV measurements around the same point (I,V) to determine the total current in the presence of BHI gives:

$$I(V) = \int_{-\infty}^{+\infty} I(\phi, V) P(\phi) d\phi. \tag{2.31}$$

The apparent reverse saturation current then be written as

$$I_o = AA^* T^2 e^{-\frac{q\phi_a}{kT}}. \tag{2.32}$$

in terms of the apparent barrier height, ϕ_a, that is due to BHI in the particular diode at the measurement temperature. Also,

$$\phi_a = \left(\bar{\phi} - \frac{q\delta^2}{2kT}\right)^2. \tag{2.33}$$

This leads to the modified form of Eq. 2.29:

$$\ln\left(\frac{I_o}{T^2}\right) + \left(\frac{q\delta}{kT}\right)^2 = -\left(\frac{q\bar{\phi}}{k}\right)\frac{1}{T} + \ln(AA^*). \tag{2.34}$$

This result is the modified Richardson's equation. Figure 2.12 compares two plots of Richardson's equation for a typical Schottky diode. The modified Richardson's plot is more linear over a wider temperature range. Figure 2.13 shows the typical barrier heights determined using the unmodified and modified Richardson's equations.

2.7.2.1 Modification of Richardson's Constant by Barrier Inhomogeneities

It is essential to mention that the values of Richardson's constant for emerging semi-conductors tend to be far lower than the theoretical values. This is usually explained

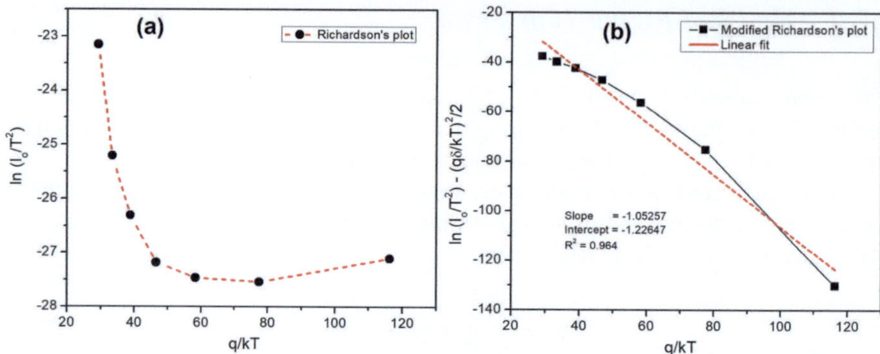

Fig. 2.12 Typical Richardson's plots for Schottky barrier diodes. In **a**, using the unmodified approach, and in **b**, using the modified approach

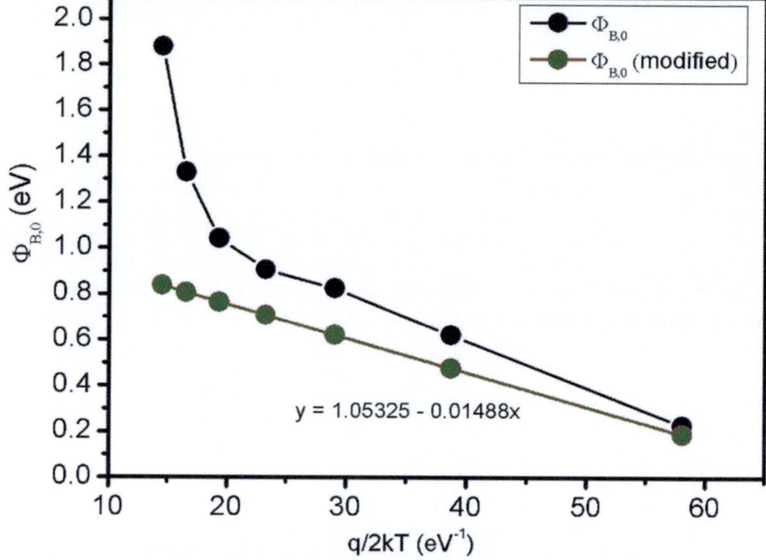

Fig. 2.13 Typical variations of the zero-bias SBH calculated using Richardson's and modified-Richardson's plots

in terms of the poor contact that these materials form in Schottky devices, resulting in inhomogeneities in their SBHs. Examples of such materials are oxide-based semiconductors such as zinc oxide, which have become the focus of active research in recent years because of their important excitonic properties [37].

These studies suggest fabrication methods that improve the contacts and accurately reflect Richardson's constants in the final devices. For instance, Zaman et al. [38] used experimental techniques similar to a previous study by Pirri et al. [39] to determine the value of Richardson's constant for Ti/4H-SiC and Mo/4H-SiC Schottky diode using forward current-voltage measurements.

Both Φ and n are temperature-dependent, with n tending to approach unity at high temperatures and increasing significantly at low temperatures. They observed this behavior and temperature sensitivity of the SBH consistently in all three diodes. The count of these patches in any SBD represents the overall effective area, denoted as A_{eff}, involved in the current transport process. Other researchers, such as Ouennoughi et al. [40], associated the temperature variations of n and Φ with interacting low-barrier patches. Typically, A_{eff} is smaller than the SBD's geometric area but plays a pivotal role in determining the total current passing through the Schottky diode, as demonstrated in Eq. 2.35:

$$I = A^{*} \cdot N \cdot A_{\mathrm{eff}} \cdot T^{2} \cdot e^{-\beta \Phi_{\mathrm{eff}}} \left(e^{\beta V_a} - 1 \right). \tag{2.35}$$

Here, $\beta = q/kT$, Φ_{eff} denotes the effective barrier height, and N stands for the total count of low barrier patches within the SBD. This count denotes a variable parameter that can change with temperature or the choice of material for the Schottky contact. $N A_{\mathrm{eff}}$ represents the total area involved in the current transport. The effective area, A_{eff}, signifies the area of a single low barrier patch and can be calculated using the following equation:

$$A_{\mathrm{eff}} = \frac{4\pi \gamma}{9\beta} \left(\frac{\eta}{V_{\mathrm{bb}}} \right)^{2/3}. \tag{2.36}$$

The terms are the band-bending V_{bb}, $\eta = \varepsilon/q N_d$ is a constant dependent on the doping density, N_d, and γ is a constant that relates the zero-bias barrier height Φ_0 to the effective barrier height i.e.:

$$\Phi_{\mathrm{eff}} = \Phi_0 - \gamma \left(\frac{V_{\mathrm{bb}}}{\eta} \right)^{1/3}. \tag{2.37}$$

Zaman et al. also incorporated the theoretical approach of Tung [36] to account for inconsistencies in A^{*} arising from nanometer-sized low-barrier patches within the diodes' high-barrier regions.

These recent findings revealed that the effective area for current transport differed significantly from the diodes' geometric area, as predicted by Tung's model. They successfully determined A^{*} values of approximately 145.39, 148.33, and 148.33 A/cm^{2}K^2 for Ti/4H-SiC (large area), Mo/4H-SiC, and Ti/4H-SiC (small area) Schottky diodes, respectively. Their research highlights the influence of various metal-SiC combinations and laboratory conditions for evaluating Richardson's constant and the effective area involved in current transport.

2.8 Schottky Barriers Using 2D Materials

Recent years have witnessed a surge in the empirical investigation of 2D phase transitions. The exploration of 2D materials had its roots in theoretical pursuits during the early 1940s that focused on attaining crucial milestones like solving the 2D Ising spin model, formulating the Hohenberg–Mermin–Wagner theorem, and identifying the Kosterlitz–Thouless transition [41]. The trajectory of progress took a significant leap forward in the 1980s, propelled by advancements in the semiconductor domain. This period yielded substantial practical insights, notably in the realm of 2D electron systems, giving rise to discoveries like the quantum and fractional quantum Hall effects.

Advances in microscopy instrumentation, including scanning tunneling microscopy and atomic force microscopy (AFM), facilitated direct observations of phenomena like surface reconstruction and phase transitions within Langmuir monolayers. The early 2000s marked a pivotal juncture with the advent of planar crystal sheets, a breakthrough that earned Geim and Novoselov the 2004 Nobel Prize in Physics and catalyzed the exploration of 2D materials. This ongoing journey spans a diverse spectrum encompassing quantum electronics, superconductivity, and condensed matter physics.

2.8.1 2D and a New Device Physics

The transition from three-dimensional (3D) to two-dimensional (2D) physics engenders distinct phenomena demanding the reconfiguration of statistical, quantum, and mechanical paradigms. The nuanced interplay of stimuli, interactions, charge transport, and phase transitions within 2D materials amplifies their sensitivity to external influences in ways profoundly different from their 3D counterparts. This distinction shapes a unique arena of physics ripe with opportunities for innovative exploration. The field of 2D materials and devices is still in its infancy.

MIS interfaces are of paramount importance in the realm of two-dimensional (2D) semiconductor devices. These interfaces are influenced by complex Fermi pinning and defect states. These interfaces are demonstrating exceptional potential. They have studied by integrating distinct 2D metal modes and 2D, monolayer semiconductors. The unique properties of van der Waals forces between these two 2D materials have reportedly been used to mitigate the effects of Fermi pinning that plagues conventional metal-semiconductor interfaces and substantial increase in SBH and near-unity ideality factor [42].

Due to the intricate layered structure of 2D semiconductors, traditional metallization methods have the potential to harm their inherent characteristics, resulting in elevated contact barriers and substantial contact resistance, R_c. Novel strategies such as utilizing van der Waals forces in metal contacts, implementing edge contacts, employing phase engineering, and integrating contact doping have been introduced and empirically validated to preempt these challenges. However, it is essential to enhance their scalability and align them more closely with industry standards and requirements [43]. In 2D materials, R_s is considered an extrinsic parameter. In such cases, it is more useful to speak of R_c, which is also an extrinsic parameter. Furthermore, many concepts, such as Richardson's equation, originally formulated for bulk materials, may still have utility in explaining the crossing of charge carriers across the Schottky barrier by thermionic emission. In atomically thin 2D heterostructures comparable to the de Broglie wavelength, van der Waals interactions dominate but the conventional TE model disintegrates.

Recent studies have addressed the need for new physics by developing comprehensive numerical models that incorporate both thermionic and field emission processes during carrier injection at the Schottky contact. These models have successfully yielded analytical IV characteristics by accounting for electrostatics and tunneling effects at the barrier. However, these numerical approaches essentially emulate 3D metal's thermionic, thermionic field-emission, and field-emission processes within the context of 2D semiconductors. Given the practical challenges associated with empirical measurements on 2D structures, advancements in this field will be bolstered by computer simulations [44].

In passing, it is necessary to mention that 3D approaches have been applied to 2D with some measure of success, such as by Riazimehr et al. [45], who used both current-voltage and capacitance-voltage measurements on graphene-silicon heterojunction photodiodes.

2.8.2 New 2D Characterization Methods

2D structures are demonstrating promising developments toward ultrathin and high-performance photovoltaic devices. However, such devices encounter formidable challenges in surpassing an external quantum efficiency (EQE) of 50% and a power conversion efficiency (PCE) of 3%. These limitations stem from the relatively inefficient separation and collection of photocarriers.

Again, fully capitalizing on van der Waals metal contacts, finding techniques to minimize interface defects, and Fermi-level pinning, gives ample future opportunities for enhancing these figures of merit. This includes techniques and materials that prevent photocarrier recombination thus enabling intrinsic and highly efficient photocarrier separation and collection. The ideal EQE value approaches the Shockley-Queisser limit [46]. 2D devices are still in their infancy, and new physics mandates new methods of electrical measurements and characterization. The recent detailed review by Pham et al. [43] indicates these dual needs clearly.

2.9   Remarks

In the realm of semiconductor devices, point-contact junctions have historically faced challenges related to repeatability, fabrication reliability, consistent operation, and characterization. Surfaces of both metals and semiconductors, despite the crystalline structure in their bulk, are defect-ridden regions due to the disruption of long-range atomic ordering at the surface, a phenomenon referred to as relaxation. Moreover, the effects of relaxation extend several lattice parameters below the surface into the bulk, playing a pivotal role in the behavior of MS devices.

In summary, the challenges associated with point-contact junctions led to the development of MS devices, which, while more robust, still grapple with surface-related complexities arising from relaxation phenomena and environmental factors like oxidation and contamination. In real devices, an ongoing challenge is the matter of parameter extraction. These factors underscore the significance of understanding surface effects in semiconductor devices.

References

1. H. Welker, Walter Schottky. Phys. Today **29**(6), 63 (1976). https://doi.org/10.1063/1.3023533
2. H. Norde, A modified I-V plot for Schottky diodes with high series resistance. J. Appl. Phys. **50**(7), 5052–5053 (1979)
3. E.H. Nicollian, J.R. Brews, *MOS (metal oxide semiconductor) physics and technology* (Wiley, New York, 1982)
4. S.K. Cheung, N.W. Cheung, Extraction of Schottky diode parameters from forward current-voltage characteristics. Appl. Phys. Lett. **49**(2), 85–87 (1986)
5. Yang Zhang, Ziyu Zhang, Bixia Lin, Fu. Zhuxi, Xu. Jin, Effects of Ag doping on the photo-luminescence of ZnO films grown on Si substrates. J. Phys. Chem. B **109**(41), 19200–19203 (2005)
6. A.H. Wilson, The theory of electronic semi-conductors.-ii. Proc. R. Soc. Lond. Ser. A Contain. Papers Math. Phys. Char. **134**(823), 277–287 (1931). The Royal Society London. https://doi.org/10.1098/rspa.1931.0196
7. B.R.A. Nijboer, On the theory of electronic semiconductors. Proc. Phys. Soc. **51**(4), 575 (1939). IOP Publishing. https://doi.org/10.1088/0959-5309/51/4/303
8. W.B. Nottingham, R.H. Good, E.W. Müller, R. Kollath, G.L. Weissler, W.P. Allis, L.B. Loeb, A. von Engel, P.F. Little, W.B. Nottingham, *Thermionic Emission* (Springer, 1956)
9. C. Kittel, P. McEuen, *Introduction to Solid State Physics* (Wiley, 2018)
10. E.H. Rhoderick, R.H. Williams, *Metal-Semiconductor Contacts* (Clarendon Press, 1988)
11. J. Frenkel, On pre-breakdown phenomena in insulators and electronic semi-conductors. Phys. Rev. **54**(8), 647 (1938)
12. M. Rudan, *Physics of Semiconductor Devices* (Springer, 2015)
13. K. Noori, F. Xuan, S.Y. Quek, Origin of contact polarity at metal-2D transition metal dichalcogenide interfaces. npj 2D Mater. Appl. **6**(1), 73 (2022)
14. A.J. Bard, A.B. Bocarsly, F.R.F. Fan, E.G. Walton, M.S. Wrighton, The concept of Fermi level pinning at semiconductor/liquid junctions. Consequences for energy conversion efficiency and selection of useful solution redox couples in solar devices. J. Amer. Chem. Soc. **102**(11), 3671–3677 (1980)

15. Ş Altındal, A. Tataroğlu, İ Dökme, Density of interface states, excess capacitance and series resistance in the metal-insulator-semiconductor (MIS) solar cells. Solar Energy Mater. Solar Cells **85**(3), 345–358 (2005). https://doi.org/10.1016/0038-1101(92)90286-L

16. A. Türüt, M. Sağlam, Determination of the density of Si-metal interface states and excess capacitance caused by them. Phys. B: Condens. Matt. **179**(4), 285–294 (1992). https://doi.org/10.1016/0921-4526(92)90628-6

17. R.H. Bube, *Photoelectronic Properties of Semiconductors* (Cambridge University Press, 1992)

18. A. Rose, *Concepts in Photoconductivity and Allied Problems* (Interscience, NJ, United States, 1963)

19. M. Brahma, M.L. Van de Put, E. Chen, M.V. Fischetti, W.G. Vandenberghe, The importance of the image forces and dielectric environment in modeling contacts to two-dimensional materials. npj 2D Mater. Appl. **7**(1), 14 (2023). https://doi.org/10.1038/s41699-023-00372-6

20. S.R. Evans, E. Deylgat, E. Chen, and W.G. Vandenberghe, *Image-Force Barrier Lowering for Two-Dimensional Materials: Direct Determination and Method of Images on a Cone Manifold*, preprint arXiv:2301.05373, 2023

21. Y. Vaknin, R. Dagan, Y. Rosenwaks, Schottky barrier height and image force lowering in monolayer MoS_2 field effect transistors. Nanomaterials **10**(12), 2346 (2020). https://doi.org/10.3390/nano10122346

22. M. Sağlam, E. Ayyildiz, A. Gümuş, A. Türüt, H. Efeoğlu, S. Tüzemen, Series resistance calculation for the Metal-Insulator-Semiconductor Schottky barrier diodes. Appl. Phys. A **62**(3), 269–273 (1996)

23. E. Ayyildiz, A. Türüt, H. Efeoğlu, S. Tüzemen, M. Sağlam, Y.K. Yoğurtçu, Effect of series resistance on the forward current-voltage characteristics of Schottky diodes in the presence of interfacial layer. Solid-State Electr. **39**(1), 83–87 (1996)

24. E. Ayyildiz, C. Temirci, B. Bati, A. Türüt, The effect of series resistance on calculation of the interface state density distribution in Schottky diodes. Int. J. Electr. **88**(6), 625–633 (2001). https://doi.org/10.1080/00207210110044396

25. R.O. Ocaya, A.G. Al-Sehemi, A. Al-Ghamdi, F. El-Tantawy, F. Yakuphanoğlu, Organic semiconductor photosensors. J. Alloys Comp. **702**, 520–530 (2017). https://doi.org/10.1016/j.jallcom.2016.12.381

26. S.M. Sze, Y. Li, K.K. Ng, *Physics of Semiconductor Devices* (Wiley, 2021)

27. E.H. Nicollian, J.R. Brews, *MOS (Metal Oxide Semiconductor) Physics and Technology* (Wiley, 2002)

28. F. Yakuphanoğlu, Analysis of interface states of metal-insulator-semiconductor photodiode with n-type silicon by conductance technique. Sens. Actuators A: Phys. **147**(1), 104–109 (2008). https://doi.org/10.1016/j.sna.2008.04.007

29. A. Türüt, N. Yalçin, M. Sağlam, Parameter extraction from non-ideal C-V characteristics of a Schottky diode with and without interfacial layer. Solid-State Electr. **35**(6), 835–841 (1992)

30. A.K. Yadav, K.X. Nguyen, Z. Hong, P. García-Fernández, P. Aguado-Puente, C.T. Nelson, S. Das, B. Prasad, D. Kwon, S. Cheema, and others, Spatially resolved steady-state negative capacitance. Nature **565**(7740), 468–471 (2019). https://doi.org/10.1038/s41586-018-0855-y

31. A. Mekki, R.O. Ocaya, A. Dere, Ahmed A. Al-Ghamdi, K. Harrabi, F. Yakuphanoğlu, New photodiodes based graphene-organic semiconductor hybrid materials. Synthetic Metals **213**, 47–56 (2016). https://doi.org/10.1016/j.synthmet.2015.12.026

32. R. Joly, S. Girod, N. Adjeroud, P. Grysan, J. Polesel-Maris, Evidence of negative capacitance and capacitance modulation by light and mechanical stimuli in Pt/ZnO/Pt Schottky junctions. Sensors **21**(6), 2253 (2021)

33. H. Durmuş, Ü. Atav, Extraction of voltage-dependent series resistance from IV characteristics of Schottky diodes. Appl. Phys. Lett. **99**(9), 093505 (2011). American Institute of Physics. https://doi.org/10.1063/1.3633116

34. R.O. Ocaya, F. Yakuphanoğlu, Ocaya-Yakuphanoğlu method for series resistance extraction and compensation of Schottky diode I-V characteristics. Measurement **186**, 110105 (2021). https://doi.org/10.1016/j.measurement.2021.110105

35. C.R. Crowell, The Richardson constant for thermionic emission in Schottky barrier diodes. Solid-State Electr. **8**(4), 395–399 (1965). https://doi.org/10.1016/0038-1101(65)90116-4

36. R.T. Tung, *The* physics and chemistry of the Schottky barrier height. Appl. Phys. Rev. **1**(1) (2014). https://doi.org/10.1063/1.4858400

37. K. Sarpatwari, O.O. Awadelkarim, M.W. Allen, S.M. Durbin, S.E. Mohney, *E*xtracting the Richardson constant: IrOx/n-ZnO Schottky diodes. Appl. Phys. Lett. **94**(24) (2009). https://doi.org/10.1063/1.3156031

38. M.Y. Zaman, D. Perrone, S. Ferrero, L. Scaltrito, M. Naretto, Evaluation of correct value of Richardson's constant by analyzing the electrical behavior of three different diodes at different temperatures. Mater. Sci. Forum **711**, 174–178. Trans Tech Publ, (2012). https://doi.org/10.4028/www.scientific.net/MSF.711.174

39. C.F. Pirri, S. Ferrero, L. Scaltrito, D. Perrone, S. Guastella, M. Furno, G. Richieri, L. Merlin, Intrinsic 4H-SiC parameters study by temperature behaviour analysis of Schottky diodes. Microelectr. Eng. **83**(1), 86–88 (2006). https://doi.org/10.1016/j.mee.2005.10.031

40. Z. Ouennoughi, S. Toumi, R. Weiss, Study of barrier inhomogeneities using I-V-T characteristics of Mo/4H-SiC Schottky diode. Physica B: Condens. Matter **456**, 176–181 (2015). https://doi.org/10.1016/j.physb.2014.08.031

41. W. Li, X. Qian, J. Li, Phase transitions in 2D materials. Nat. Rev. Mater. **6**(9), 829–846 (2021). https://doi.org/10.1038/s41578-021-00304-0

42. X. Zhang, B. Liu, L. Gao, H. Yu, X. Liu, J. Du, J. Xiao, Y. Liu, L. Gu, Q. Liao et al., Near-ideal van der Waals rectifiers based on all-two-dimensional Schottky junctions. Nat. Commun. **12**(1), 1522 (2021). https://doi.org/10.1038/s41467-021-21861-6

43. P.V. Pham, S.C. Bodepudi, K. Shehzad, Y. Liu, Y. Xu, B. Yu, X. Duan, 2D heterostructures for ubiquitous electronics and optoelectronics: principles, opportunities, and challenges. Chem. Rev. **122**(6), 6514–6613 (2022). https://doi.org/10.1021/acs.chemrev.1c00735

44. A. Tunga, Z. Zhao, A. Shukla, W. Zhu, S. Rakheja, Physics-based modeling and validation of 2D Schottky barrier field-effect transistors (2023). arXiv:2307.04851

45. S. Riazimehr, M. Belete, S. Kataria, O. Engström, M.C. Lemme, Capacitance-voltage (C-V) characterization of graphene-silicon heterojunction photodiodes. Adv. Opt. Mater. **8**(13), 2000169 (2020). https://doi.org/10.1002/adom.202000169

46. H. Wang, W. Wang, Y. Zhong, D. Li, Z. Li, X. Xu, X. Song, Y. Chen, P. Huang, A. Mei, and others, Approaching the external quantum efficiency limit in 2D photovoltaic devices. Adv. Mater. **34**(39), 2206122 (2022). https://doi.org/10.1002/adma.202206122

Chapter 3
Contemporary Parameter Extraction Methods

Abstract This chapter reviews the currently used methods of parameter extraction in the literature. Since the discovery of MS devices, several methods have been suggested, with varying complexity and utility. Here, we present a cross-section of the most recent electrical methods devised over the last decade. The methods are divided into three broad categories. These are current-voltage methods, capacitance-voltage (impedance) methods, and transport methods.

3.1 Classification of Extraction Methods

There is an abundance of the reported methods that attempt to address the challenging problem of extracting the static parameters of variously configured MS diodes. The impetus for accurate parameters is not only the characterization of new devices but also the need to develop correct simulation models that extend new applications [1]. Simulations are becoming crucial owing to the growing difficulties in physically probing at the unfathomable levels of miniaturization, without introducing measurement errors. These methods generally assume non-degenerate semiconductors. That is, $E_F \geq 3kT$ than the band edges, preventing metal-like behavior [2, 3] and giving discrete energy levels within the band gap. This can be obtained by moderately doping the semiconductors such that the dopants are non-interacting due to their wide separation.

Over the last decade, many measurement methods of varying complexity and accuracy have been devised. These methods can be categorized into four board groups: current-voltage (I-V), activation energy, capacitance-voltage (C-V), and photoelectric measurements [4]. Although these measurements often rely on simplifying assumptions, they provide reasonable indications of expected parameter ranges and variations.

The C-V method, for instance, is commonly performed in reverse bias using a high-frequency excitation signal, assuming that the diode will not exhibit a low-voltage resonance peak [2]. This peak is believed to be caused by interfacial charges that follow the alternating current signal and contribute to measured capacitance at

R. Ocaya, *Extraction of Semiconductor Diode Parameters*,
https://doi.org/10.1007/978-3-031-48847-4_3

frequencies below 1 MHz [5, 6]. The specific assumption is that the capacitance of interest arises solely from the space charge [2].

In the I-V method, it is crucial to consider the effect of the diode's forward resistance. The existing I-V extraction methods may be subdivided further into three categories: simplistic models based on TE theory, the physical analytical models providing closed-form solutions of transport equations, and numerical models based on Poisson's equation, drift-diffusion, and continuity equations [3, 7]. It has been shown in [3] that many engineering methods for processing I-V curves to extract static parameters impose restrictive assumptions on acceptable parameter values. For example, Norde's and other methods [8–10] are limited to R_s in the range of 10–20 Ω, which is not applicable to all Schottky diodes. Additionally, these methods may have features that directly and adversely impact their accuracies. It is worth noting that few existing methods address the issue of resistance [3]. The classical approach involves determining Φ_b and n in a low-current mode where the effect of resistance is negligible. However, this approach can be problematic when dealing with devices that require higher characterization currents.

In this chapter the foremost approaches for MS parameter extraction are discussed but with more emphasis on newer methods. By newer we mean those devised over the last decade. In order to maintain scope, we will focus on the ones that prioritize R_s and interface trap density due to the limitations that these parameters impose on MS device performance.

3.2 Current-Voltage Methods

The foremost and almost de facto I-V characterization methods are Norde's method and its derivatives, and the Cheung-Cheung method [11, 12]. The latter method, in particular, is the more recent of the two and is generally considered the benchmark against which subsequent methods up to the present have been evaluated. Both these methods have withstood the test of time despite their less-than-perfect performances. Purely theoretical error analyses of these methods have been largely absent, and such new methods have relied on empirical fits on some, often narrowly-scoped experimental data to vouch for their validity.

Although both of these popular I-V methods are generally successful in extracting parameters under many device conditions and configurations, they are limited in a number of crucial ways. For instance, they are impacted significantly by the chosen range of bias, show sensitivity to large device R_s, and report high variance in the extracted parameters, usually under heavy and rather unrealistic approximations. The result is that the extracted values often conflict even within the methods themselves. For one, the majority of I-V methods in existence are semilog approximations reliant on discarding the exponential terms. This requires the net applied bias in Eq. 2.26 to be much greater than $3kT$.

Furthermore, in light of the complexity of the TE equation, rigorous error analyses of the various methods have been conspicuously absent over the years, barring side-by-side analysis of their efficacy on the same I-V data pertaining to a hypothetical device. This makes their impact truly difficult to ascertain. We review the better-known I-V methods, beginning with Norde's method and its extensions, and the Cheung-Cheung method, for their obvious importance.

3.2.1 The Norde Method

The method presented in 1979 by Herman Norde remains an important mathematical exercise that attempts to address the extraction of MS device parameters from a highly non-linear equation that arises from significant R_s [11].

The method is still in use under various modifications, due to the quickly realized limitations of the original form. For instance, its accuracy diminishes when there are notable effects on the forward current, such as recombination, minority-carrier injection, tunneling, or when the Schottky barrier height exhibits bias [13]. Also, as Norde himself noted, it is often difficult to ascertain whether the source of R_s which it eventually treats is a true internal resistance or a resistance from a non-Ohmic source, such as a device back-contact effect.

To see the essence of the method, when R_s is set to zero in Eq. 2.26, the TE equation takes the ideal form

$$I = AA^*T^2 e^{-\beta\Phi}\left(e^{\beta V} - 1\right), \tag{3.1}$$

where $\beta = q/kT$. Then, under the condition that $\beta V >> 1$,

$$I \approx AA^*T^2 e^{-\beta\Phi} e^{\beta V}. \tag{3.2}$$

The plot of $\ln(I)$ against applied bias V is ideally a straight line with an intercept (at $V = 0$) that gives the reverse saturation current at a given temperature (T), for a specific diode geometry (AA^*). It is then evident that

$$\frac{1}{\beta}\ln\left(\frac{I}{AA^*T^2}\right) \approx V - \Phi \tag{3.3}$$

Norde suggested that the issue arising from R_s can be handled meaningfully if one defines an essentially linear function $F(V)$; additionally, in the presence of any non-linearity, the function should ideally express any deviations from linearity when graphically plotted against V.

Such a function is

$$F(V) = \frac{V}{2} - \frac{1}{\beta}\ln\left(\frac{I}{AA^*T^2}\right). \tag{3.4}$$

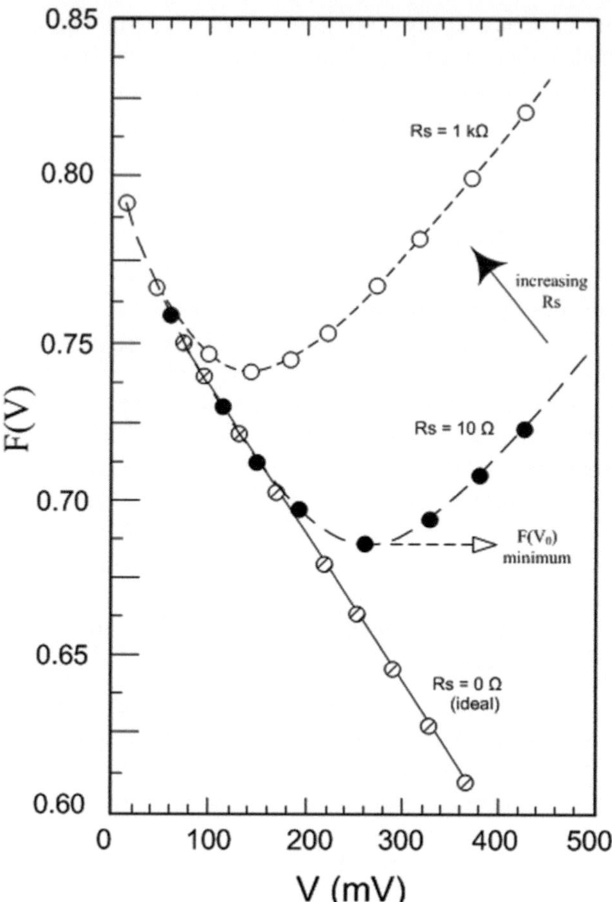

Fig. 3.1 Hypothetical plots of the Norde function $F(V)$ versus bias V for typical Schottky diodes. The plots range from the ideal diode with $R_s = 0$, to relatively large R_s. The $R_s > 0$ plots have a minimum point at a specific applied bias, denoted V_0. This point is then used to determine $\Phi \approx 0.79$ eV

Substituting the logarithm term in Eq. 3.4 with Eq. 3.3 gives

$$F(V) = -\frac{V}{2} + \Phi. \tag{3.5}$$

The intercept of the $F(V)$ is then the barrier height, Φ. Also, Φ may be calculated at any $(V, F(V))$ point for the ideal barrier diode. Figure 3.1 shows the Norde plot for $R_s = 0$. Implicit in Norde's argument is that the current is essentially a constant in the neighborhood of measurement. Therefore, he took the normal derivative of $F(V)$ w.r.t to V, which equals $\frac{1}{2}$. In reality, I depends on the bias point at which it is measured so taking the partial derivative is probably more meaningful.

Norde postulated deviations from linearity due to non-zero R_s by modifying the bias term in Eq. 3.3 to $(V\text{-}IR_s)$ while leaving the *definition* of Eq. 3.4 unchanged s.t.

$$F(V) = -\frac{V}{2} + \Phi + IR_s, \qquad (3.6)$$

and under the condition that $kT/q << V << IR_s$. It turns out that $F(V)$ expresses a well-defined minimum point in the presence of R_s-induced nonlinearity, as shown in Fig. 3.1 for $R_s > 0$. In applying this condition, Norde assumed that Φ is independent of applied bias, an assumption that is now known to be generally untrue. Consequently,

$$\frac{dF}{dV} = -\frac{1}{2} + R_s\left(\frac{dI}{dV}\right). \qquad (3.7)$$

This equation allows R_s to be determined at the minimum point by first writing Eq. 3.2 in the presence of R_s i.e.:

$$I \approx I_s e^{\beta(V-IR_s)}, \qquad (3.8)$$

which leads to

$$\frac{dI}{dV} \approx \beta I\left[1 - R_s\frac{dI}{dV}\right] \implies \frac{dI}{dV} \approx \frac{\beta I}{1 + R_s\beta I}. \qquad (3.9)$$

Substituting this result into Eq. 3.7 gives

$$\frac{dF}{dV} = -\frac{1}{2} + \frac{R_s\beta I}{1 + R_s\beta I} = \frac{-1 + R_s\beta I}{2(1 + R_s\beta I)}. \qquad (3.10)$$

The minimum point of the $F(V)$ occurs when $F'(V) = 0$, at the I-V graphical minimum point designated (I_0, V_0), hence

$$R_s = \frac{1}{\beta I_0} = \frac{kT}{qI_0}. \qquad (3.11)$$

As a result, the barrier height can be found at the minimum point V_0 using the foregoing equations:

$$\Phi = F(V_0) + \frac{1}{2}V_0 - \frac{kT}{q}. \qquad (3.12)$$

Norde had primarily considered ideal diodes ($n = 1$) and postulated a method that relies on two characteristics at different temperatures to completely characterize the diode. Within a few years, several others, such as Sato and Yasumura [14], Bohlin [15] presented extensions of the Norde plot methodology to non-ideal diodes

($n > 1$) for determining R_s, Φ, and n. Sato and Yasumura's method was based on two temperatures, while Bolin's used a single current-voltage (I-V) measurement [15].

The first of the two extensions is more infrequently used compared to the second because of its reliance on two temperatures. Both methods are prefixed with the name Norde since they are derived from it. They both consider non-ideal diodes with non-zero R_s i.e., they modify Eq. 3.2 to

$$I \approx AA^*T^2 e^{-\beta\Phi} e^{\beta(V - IR_s)/n}. \tag{3.13}$$

They are both described briefly below.

3.2.2   Norde–Sato–Yasumura Method

Sato and Yasumura define their $F(V)$ function as

$$F(V) = \left(\frac{1}{2} - \frac{1}{n}\right)V + \Phi + \frac{IR_s}{n}. \tag{3.14}$$

Following the above steps for Norde's method, their condition for minima in $F(V)$ is then

$$\frac{dF}{dV} = \frac{n - 2 + R_s\beta I}{2(n + R_s\beta I)}. \tag{3.15}$$

The condition for minima is then $F'(V) = 0$, so that at the minimum point (I_0, V_0):

$$R_s = \frac{2 - n}{\beta I_0}, \tag{3.16}$$

and

$$\Phi = F(V_0) + \left(\frac{1}{2} - \frac{1}{n}\right)V_0 - \left(\frac{1}{n} - 1\right)\frac{kT}{q}. \tag{3.17}$$

Therefore, the Sato–Yasumura extension reduces to Norde's method when $n = 1$. It is soon clear that to determine n, R_s, and Φ, at least two sets of measurements at two different temperatures T_1 and T_s are needed. The general assumption from such measurements is that these parameters are constant over temperature. Again, this is not generally true, particularly when the two temperatures are significantly different, as the case would be if there is to be any expectation of improving the measurement uncertainties.

3.2.3 *Norde–Bohlin Method*

The extension devised by Bohlin averts the need for at least two different temperatures of measurement of the IV data. The equivalent functions introduced by Bohlin have similar terms, but with the exception of the presence of some arbitrarily chosen constant, $\gamma > n$ i.e.:

$$F(V) = \left(\frac{1}{\gamma} - \frac{1}{n}\right)V + \Phi + \frac{IR_s}{n}. \tag{3.18}$$

Following the above steps for Norde's method, their condition for minima in $F(V)$ is then

$$\frac{dF}{dV} = \frac{n - \gamma + R_s\beta I}{\gamma(n + R_s\beta I)}. \tag{3.19}$$

The condition for minima is then $F'(V) = 0$, so that at the minimum point (I_0, V_0):

$$R_s = \frac{\gamma - n}{\beta I_0}, \tag{3.20}$$

and

$$\Phi = F(V_0) + \left(\frac{1}{n} - \frac{1}{\gamma}\right)V_0 - \left(\frac{\gamma}{n} - 1\right)\frac{kT}{q}. \tag{3.21}$$

It is then necessary to use two different γ, through which the parameters may be extracted. Like the first extension to Norde's method, Bohlin's method also reduces to Norde's method when $n = 1$.

3.2.4 *The Cheung-Cheung Method*

In their 1986 letter, Cheung and Cheung postulated two linear functions whereby n, R_s, and Φ could be obtained graphically from a single I-V curve, in contrast to Norde's method five years prior. They validated their method on thermally annealed W/GaAs diodes. Their motivation was an alternate approach to Norde, and derivatives of Norde's method that considered non-ideal diodes, e.g., by Sato and Yasumura [14].

The central tenet of their method [12] involves differentiating the current density $(J = I/A)$ form of Eq. 2.26 at the temperature T, and defining a function $H(J)$:

$$\frac{dV}{d\ln I} = R_s I + \frac{nkT}{q} \tag{3.22}$$

and

$$H(I) = R_s I + n\Phi_b = V - \frac{nkT}{q}\ln\left(\frac{I}{AA^*T^2}\right), \tag{3.23}$$

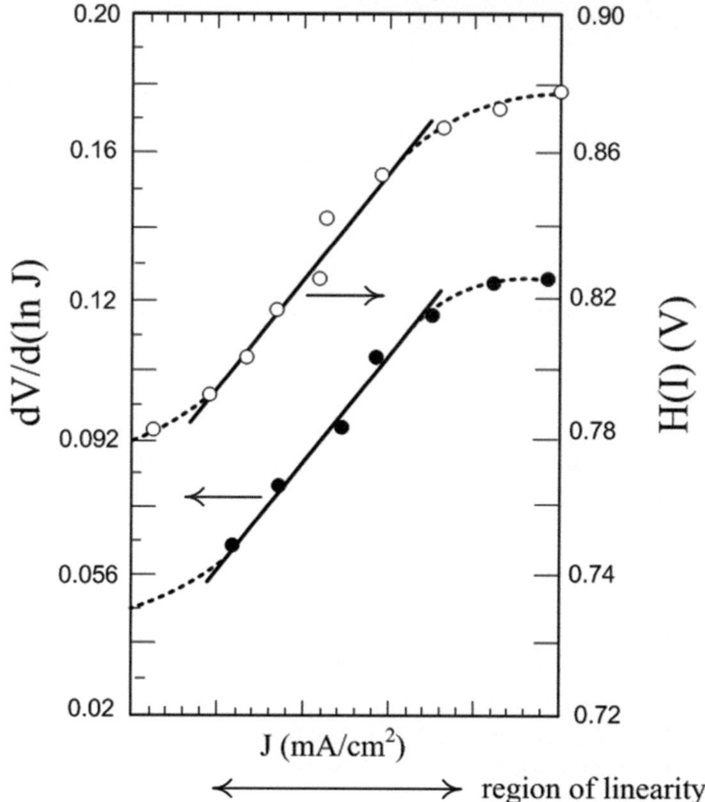

Fig. 3.2 Hypothetical plot of the Cheung-Cheung functions of a typical Schottky diode with relatively large R_s. The gradients of the functions are ideally identical and equal to R_s. The plots indicate that in such a case, the range of bias and current density that gives good linearity is narrow

with the symbols as previously defined in the TE equation. The functions $dV/d\ln(I)$ and $H(I)$, when plotted against the diode current I, are generally linear and allow two estimates of R_s which should converge. That is, the two functions are expected to have the same gradient R_s.

Consequently, the intercept of the first plot leads to an estimate of n, and Φ_b arises from the intercept of the second plot. However, the condition imposed by Cheung and Cheung, though not explicitly stated, is that all the extracted parameters R_s are fundamentally constant over the applied bias. Figure 3.2 shows the typical plots of the Cheung-Cheung functions. The plots suggest that the method's validity is restricted to the low and moderate-bias regions.

However, despite the enduring popularity of the Cheung-Cheung method which is evidently on account of its inherent simplicity and relative parameter extraction speed, almost all subsequent experiments on Schottky diodes that involve extracting these parameters in different regions of bias showed, and continue to show, that these parameters are anything but constant over bias. Moreover, this is without even considering the effects of other ambient conditions such as illumination [16], temperature, and so on, which are now known to impact these parameters [17, 18]. Furthermore, devices for which the method is successful typically have R_s values that are much less than 100 Ω such that $IR_s \ll V$ in Eq. 2.26.

These implicit assumptions have the net result of averaging R_s to a constant value. The focus on this expectation highlights an important limitation of the method that becomes apparent in devices with large R_s, with the two functions often giving substantially different R_s. This could be due to the fact that when the two functions are plotted, only the visually linear and narrow regions of bias are used to determine R_s, inadvertently introducing a subjective element to the extracted parameters. In short, the Cheung-Cheung method suffers from the significant nonlinearity introduced by the term IR_s. The recent method of Ocaya and Yakuphanoğlu, discussed in Chap. 6, averts these issues.

Other noteworthy methods have been suggested subsequent to the above methods. Mikhelashvili et al. [19] proposed two parameter extraction methods for room temperature for Schottky diodes that show bias-dependent barrier heights. These diodes can be in a single or back-to-back configuration.

Figure 3.3 depicts the back-to-back device configuration. In the case of a single rectifying junction, one of the diodes is absent from the diagram. In reality, it tends to be present owing to the fact that realistic junctions are not purely ohmic. Their method acknowledges the significant impact of R_s on the other parameters.

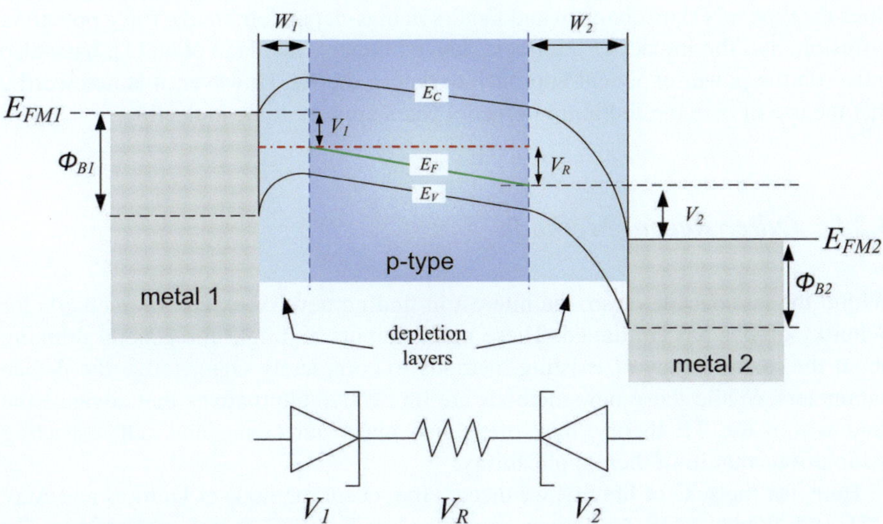

Fig. 3.3 Depiction of the energy band diagram of a Schottky diode that effectively acts like two barriers in a back-to-back configuration. Such diodes are often created, even unwittingly, during fabrication. Adapter from Kasani [20]

The technique relies on I-V measurements conducted under various bias polarities, illumination intensities, and voltage-dependent differential slope-voltage character- istics. While their technique has been experimentally validated for different Schottky diode configurations, their prior work suggested its applicability under assumptions of both linear and nonlinear R_s [21]. Their use of a power exponent parameter, α:

$$\alpha = \frac{d(\ln I)}{d(\ln V)},\tag{3.24}$$

informs which one of the two sub-methods applies to the particular device. R_s mod- ifies the voltage drop at the junction, s.t.

$$V = \frac{nkT}{q}\ln\left(\frac{I}{I_o}\right) + A_0 I^{1/\beta} + I R_s,\tag{3.25}$$

where $\beta > 1$ is a nonlinear contribution, and A_0 is a constant that depends on the intrinsic parameters of the semiconductor materials and temperature. The other vari- ables are as given in Eq. 2.27. They established that R_s influences the current flow mechanism at large bias, although the value of R_s itself is almost independent of bias under certain conditions.

These conclusions are counter-intuitive and directly contrast recent work by many others, for instance, Ocaya et al. [17]. Arguably, despite its complexity, the method eases the determination of Φ without requiring specific knowledge of interface states or insulator layers at the metal-semiconductor (MS) boundaries. Additionally, it iden- tifies the type of TE mechanism and factors in bias-dependent image force potential, diffusion, and the impact of interfacial layers. Others, like Hao et al. [1], have also relied on the power exponent approach to determine R_s. However, it is noteworthy that the use of α in the literature in recent years appears to be in decline.

3.2.5   Other Recent Methods

Within the last decade or so, the interest in finding new IV extraction methods for Schottky diodes has continued. These new attempts highlight the general thinking about the insufficiency of existing methods to completely characterize the device parameters. While these new methods are innovative alternatives that advance the field within the TE theory, they often also make narrowing and self-restricting assumptions that limit their applicability.

Here, for the sake of brevity, we discuss the recent methods of Durmuş and Atav [22], and Nouchi [23]. As already mentioned in Sect. 3.2, it is important to point out from the onset that these new methods are theoretical in nature and rely on *hypothetical* data for validation. Therefore, it becomes difficult to conduct a side-by- side comparison of their relative efficacies.

Durmuş and Atav begin by making a number of assumptions in their method, besides the validity of thermionic emission theory. Specifically, that:

- the ideality factor is a smooth function of the applied bias, V, i.e.:

$$\frac{1}{n} = 1 - \frac{\partial \Phi}{\partial V}, \tag{3.26}$$

s.t. logarithm of the ideality factor i.e., ln (n) is a weak function of V, and
- the effective barrier height, Φ is linearly dependent on V i.e.:

$$\Phi = \Phi_0 + \xi V, \tag{3.27}$$

where Φ_0 is the zero-bias barrier height and ξ is a constant voltage-coefficient of barrier height.
- There is no temperature dependence.

The first implies that $dn/dV \to 0$. Taken together, these assumptions imply that $\xi \to 0$, since $n = 1/(1 - \xi)$.

For analytical convenience, Durmuş and Atav adopt the power-law exponent definition used by Mikhelashvili and Eisenstein [19, 21] i.e.:

$$\alpha = \frac{d(\ln I)}{d(\ln V)}, \quad \gamma = \frac{d(\ln \alpha)}{d(\ln V)}. \tag{3.28}$$

Consequently, in Schottky's TE formula Eq. 2.26, their diode current expression becomes

$$I = I_0 \exp\left(\frac{V - IR_s}{n\beta}\right)\left(1 - \exp\left(-\frac{V - IR_s}{\beta}\right)\right) \tag{3.29}$$

where $\beta = kT/q$ and $I_0 = AA^*T^2 \exp(-\Phi_0/\beta)$. Defining Λ as $(V - IR_s)/\beta$ and using Eq. 3.28, then gives

$$d\Lambda = \frac{\alpha}{V}dV. \tag{3.30}$$

Also, for $V > 3\beta$, Eq. 3.29 becomes

$$I = I_0 e^{\Lambda}. \tag{3.31}$$

The derivative of this last equation w.r.t. V is,

$$\frac{\alpha}{V}(n\beta + IR) = 1 - I\frac{dR_s}{dV} - (V - IR_s)\frac{d(\ln n)}{dV}. \tag{3.32}$$

3.2.6 Determining R_s and I_o

At a given experimental (I, V) point, the term Λ is essentially fixed s.t. Eq. 3.32 takes the linear first-order ordinary differential equation (ODE) form:

$$\frac{dR_s}{dV} - \frac{\alpha}{V}\left(\frac{1+\Lambda}{\Lambda}\right)R_s + \frac{\alpha-\Lambda}{\Lambda I} = 0.$$

(3.33)

They construct a solution of this ODE of the form:

$$R_s(V) = \frac{\Lambda}{T}\int_{V_0}^{V}\frac{\Lambda-\alpha}{\Lambda^2}dV + R_0\frac{\Lambda}{\Lambda_0}e^{-(\Lambda-\Lambda_0)},$$

(3.34)

where R_0 is $R_s(0)$.

The solution $R_s(V)$ denotes the instantaneous value of R_s at the experimental point that is normalized for the saturation current at Λ_0 s.t. $dR_s/d\Lambda_0 = 0$. They subsequently assume a special case of their derivations in which both R_s and n are constants, and like Mikhelashvili and Einsenstein, arrive at:

$$R_s(V) = \left(\frac{1-\gamma}{\alpha^2}\right)\frac{V}{I}.$$

(3.35)

Even though they did not point it out, the instantaneous value of R_s according to their method is a modified version of the resistance at the measurement point. If the latter is (I_m, V_m), with the subscripts denoting a measured value, then

$$R_s(V) = \left(\frac{1-\gamma}{\alpha^2}\right)R_m = \delta R_m,$$

(3.36)

where $R_m = V_m/I_m$, for some exponent-factor determined constant, $\delta = (1-\gamma)/\alpha^2$.

This last result is extremely interesting and important, since it relates the methods of Durmuş and Atav, and Mikhelashvili and Einsenstein, to the new method and potentially unifying method of Ocaya and Yakuphanoğlu [17] which uses a completely different analytical approach based on symmetry. Real experiments, discussed later, show that there is a bias dependence of R_s on V, suggesting that both α and γ are sensitive to the region of bias in which they are determined.

As it is described, using Eq. 3.31, one obtains

$$\ln I = \ln I_o + \Lambda.$$

(3.37)

Therefore, by obtaining an empirical plot of $\ln I$ versus Λ, the reverse saturation current I_o can be determined from the intercept.

Finally, Durmuş and Atav themselves identified potential issues with their own method. Specifically, they found that the method is unsuitable for cases that conflict

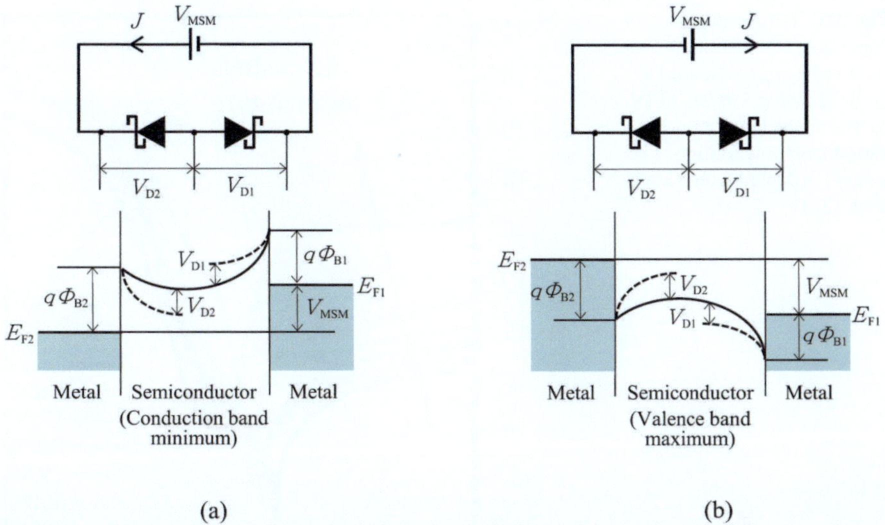

Fig. 3.4 The diagrammatic representation of two back-to-back Schottky diodes used to illustrate the parameter extraction method. Adapted from Nouchi [23]

with the assumptions of weak-voltage dependence of n. Furthermore, they noted that R_s versus V plots displayed unusual peaking behavior. It remains unclear whether this phenomenon is reported when analyzing real devices using their approach or is just seen in their hypothetical data.

Nouchi [23] developed a method for extracting parameters from the IV characteristics of metal-semiconductor-metal (MSM) diodes with asymmetric Schottky barrier heights within the TE model. When two barriers exist, the smaller barrier can usually be ignored. The MSM diode structures depicted in Fig. 3.4 were chosen because they eliminate the need for an additional MS interface, which often introduces a charge-injection barrier. Nouchi began by writing the Schottky TE Eq. 2.26 in current density form in terms of the saturation current density and the instantaneous current density for each diode i.e.:

$$V_{D1} = -n\beta \ln\left(1 - \frac{J}{J_{S1}}\right), \quad V_{D2} = n\beta \ln\left(1 + \frac{J}{J_{S2}}\right), \qquad (3.38)$$

where $\beta = kT/q$.

By analyzing the first-order derivative of the I-V curve and detecting local maxima, the method allows for the extraction of parameters such as the zero-bias barrier heights of both MS interfaces, the series resistance of the MSM diode, and the effective ideality factor for the MS diode with the higher barrier. The maximum is the point of intersection of the two JV curves, under the condition that the applied biases are both greater than $n\beta$. The method does not take into account image-force effects which results in barrier height lowering.

Fig. 3.5 Temperature-dependent I-V characteristics (IVT) of a typical Schottky diode showing a series of IV curves as the temperature is varied from low to high values. Adapted from Ocaya et al. [25]

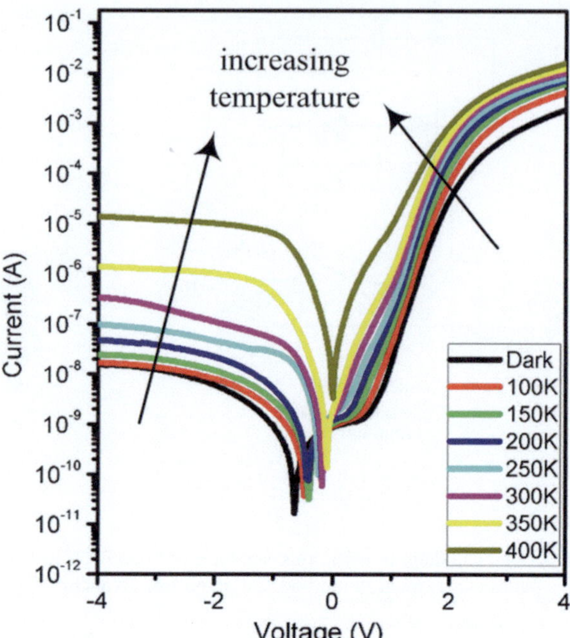

3.3   Current-Voltage-Temperature Methods

Temperature is a crucial ambient variable largely understated in most literature, where isothermal measurements are typically presumed. The IV performances of Schottky barrier diodes are strongly temperature-dependent. In many contexts, when the temperature is referred to it is usually in the unrelated context of device fabrication or thermal degradation [24]. Figure 3.5 shows the typical I-V characteristics under the effects of increasing temperature in both forward and reverse bias. The reverse currents are also seen to depend strongly on changes in temperature.

The impact of temperature in modifying the IV characteristics deserves a separate discussion and, consequently, we now examine the IV-temperature (IVT) methods. A more detailed discussion of IVT methods is done in Chap. 5, due to the potential for more accurate extraction compared to other IV methods.

3.4   Capacitance-Voltage Methods

Device parameter extraction methods based on capacitance are collectively referred to as AC-impedance or capacitance-voltage-frequency (CVF), or simply as C-V techniques. The measurements of device parameters related to capacitance therefore require alternating current (AC) excitation over applied bias and frequency. This

Fig. 3.6 Typical representation of a metal-oxide capacitor (MOS)

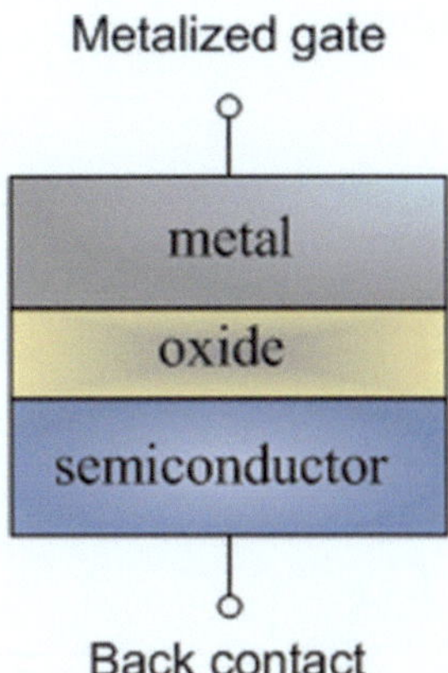

is because the inherent dielectric materials of capacitors are insulators. Therefore, such devices essentially block direct currents (DC). Schottky diodes are special-case metal-oxide semiconductor (MOS) capacitors.

The measurements of the C-V responses of MOS capacitors have important advantages such as facilitating the determination of the interface state density, D_{it}, and the built-in potential, V_{bi}, and leading to the categorizing of the energy distribution of defect states, E_{ss}, in the devices. These parameters exhibit strong bias and frequency dependence.

The essence of the C-V method is Eq. 2.18. Methods have been devised that permit the extraction of many of the parameters of the MS barrier that manifest in the capacitance. For instance, in MS devices, it includes the p- or n-type substrate, the interfacial/oxide layers, the depletion and junction capacitance, the series resistance, the interface state density, the non-ionized acceptor (or donor in n-type) density, and other parameters [26]. Figure 3.6 depicts the typical representation of the MS barrier diode.

However, CVF measurements cannot determine n outright and are notoriously divergent from the alternative methods in calculated Φ and R_s values. This divergence can be attributed to the influences of R_s and the depletion capacitance within the device. Also, for back-to-back Schottky devices, determining which device is contributing to the capacitance, and under which specific bias condition, can be challenging. To mitigate these discrepancies and portray a more accurate picture of the device's AC response, an adjusted capacitance (C_a) and conductance (G_a) are derived from circuit networks [27] in terms of measured capacitance and conductance of C_m and G_m, respectively, as follows [28, 29]:

$$C_a = \left[\frac{G_m^2 + (\omega C_m)^2}{a^2 + (\omega C_m)^2} \right] C_m, \qquad (3.39)$$

and

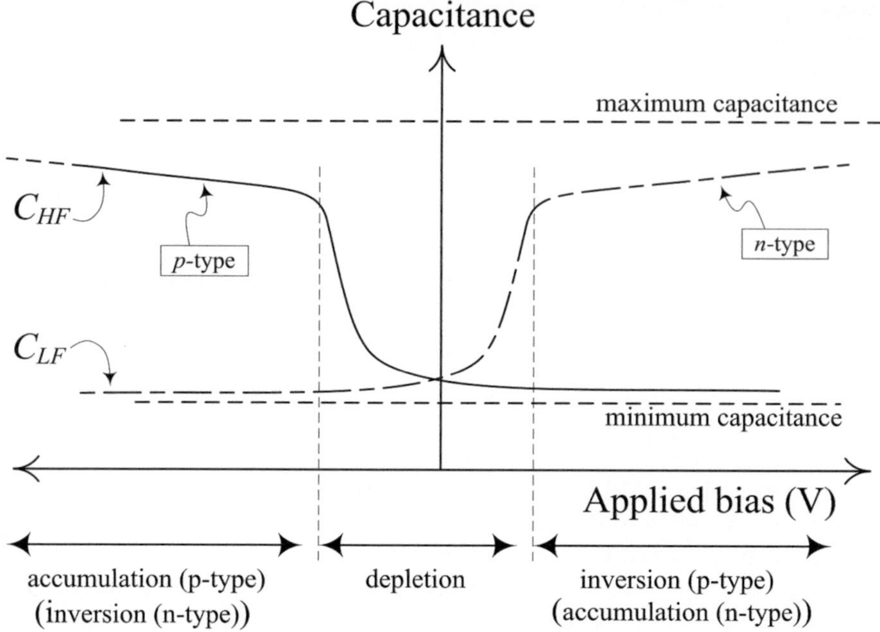

Fig. 3.7 Typical regions for capacitance measurement on Schottky diode. The diagram shows that the measurements on p-type are essentially reversed for an n-type device

$$G_a = \left[\frac{G_m^2 + (\omega C_m)^2}{a^2 + (\omega C_m)^2} \right] a, \tag{3.40}$$

where $a = G_m - [G_m^2 + (\omega C_m)^2] R_s$. The plot of Eq. 2.18 is then done using the adjusted capacitance, C_a, in place of the variable C.

Figure 3.7 shows the regions of capacitance measurements in the MOS capacitor model of a Schottky diode. The adjacent bias regions bordering the depletion region are referred to as either weak accumulation or weak inversion. Consequently, strong accumulation and strong inversions are the regions that are far from the respective boundary with the depletion region.

3.4.1   Series Resistance

A more accurate reflection of R_s of a Schottky barrier diode in the C-V method is calculated at high frequency, usually 1 MHz, in the strong accumulation region [27]. The effective circuit admittance at high frequencies can be expressed in terms of the compensated capacitance and conductance as:

$$Y_a = G_a + j\omega C_a. \tag{3.41}$$

Consequently, R_s is found using:

$$R_s = \frac{G_a}{G_a^2 + (\omega C_a)^2}.$$ (3.42)

Figure 3.8 shows a series of typical plots depicting the variation of R_s at different measurement frequencies. It is worth noting that in the literature, such plots typically exhibit a pronounced frequency dependency. Particularly at high reverse bias, the calculated R_s values using the C-V method tend to exhibit greater differentiation among themselves.

Surprisingly, as the reverse bias is increased, these R_s values generally decrease, which presents a counterintuitive aspect and underscores an inherent limitation of the C-V method. In this specific case, the two values of R_s from the Cheung-Cheung method were 143 and 146 Ω, whereas the C-V-determined R_s ranged from 68 Ω (at 1 MHz) to 149 Ω (100 kHz) at a forward bias of 2.0V. This bias point was chosen for comparison because the Cheung-Cheung functions showed linearity over a narrow range of 1.0–2.2 V in this specific case. However, R_s at 10 kHz amounts to 520 Ω at 2.0 V.

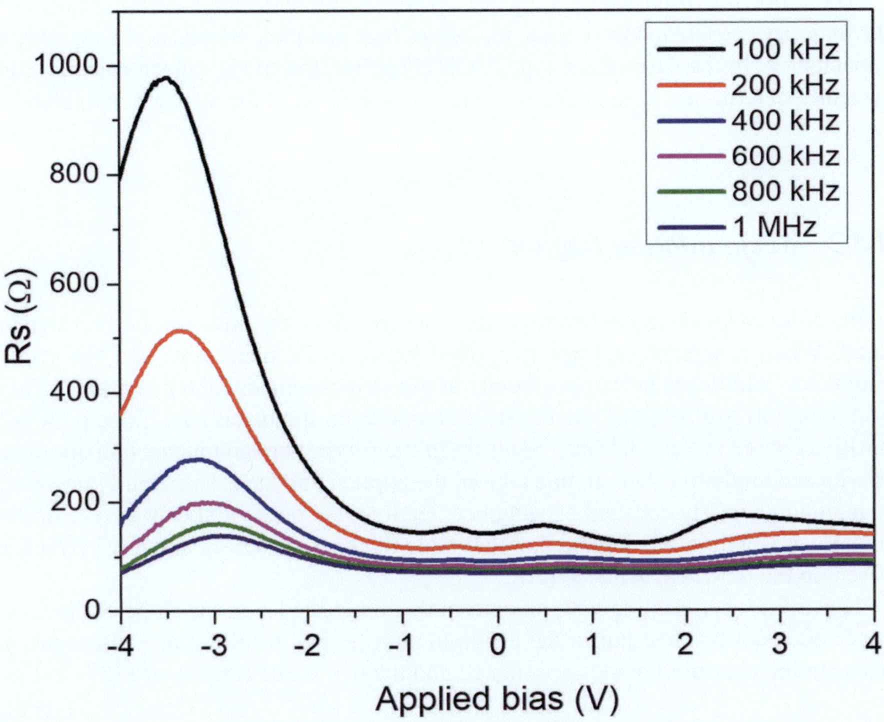

Fig. 3.8 Typical plots of series resistance versus applied bias and frequency at 300 K [25]. The plots show a strong dependence on frequency in reverse bias

At the same time, the plot shows several interesting peaking phenomena at 10 kHz measurement frequency that were not commented on in the source article. We may propose that this is due to a hitherto unsuspected negative resistance. Cheung-Cheung's method, because it essentially forces R_s to assume an averaged value of the range of linearity, would fail to detect peculiarities such as negative differential conductance (NDC). Negative This phenomenon requires further study. NDC has, in fact, recently been reported along with its possible origin, in an application of the new method by Ocaya et al. [30]. Such recent work suggests that the phenomenon may be more widespread than was previously acknowledged owing to the limitations of the analysis method.

However, when examining the forward bias in recently reported results where a number of different methods were used, it was found that the C-V method generally aligns with the newly proposed IV method by Ocaya and Yakuphanoğlu [17], which is discussed in a later chapter. In this case, the C-V method displays a relatively weaker dependence on the applied bias. Furthermore, the new method aligns intuitively with the expected behavior of R_s under reverse bias conditions. Despite the various compensations that are done on the measured variables, the reported internal parameters of the Schottky diodes can differ substantially from those obtained using I-V methods.

The following discussion pertains to a p-type semiconductor, which has holes as the majority carrier, and the unique regions of bias and their effects on the measured capacitance. In the case of a p-type MOS capacitor, the oxide capacitance, C_{ox}, is typically determined in the depletion region as well as in the strong accumulation region.

3.4.2   Accumulation Region

In the absence of an applied voltage, the holes reside in the bulk within the valence band. When a negative voltage is applied between the metal gate and the semi-conductor, additional holes accumulate at the oxide-semiconductor interface. This accumulation is a result of the negative charge from the metal gate inducing a net positive charge at the interface, resulting in the p-type semiconductor transitioning into an accumulation state. In this region, the applied voltage is sufficiently negative to maintain a nearly constant capacitance, leading to a nearly flat C-V curve. Additionally, it is within this regime that the oxide thickness can be accurately extracted based on the oxide capacitance [26].

However, when dealing with an extremely thin oxide layer, the slope of the C-V curve does not become flat in the accumulation region, resulting in a discrepancy between the measured oxide capacitance and the true oxide capacitance [27, 31].

3.4.3 Depletion Region

When a positive voltage is applied between the gate and the semiconductor, the majority carriers are displaced from the semiconductor-oxide interface, forming a depletion region at the semiconductor side of the interface. Consequently, this region of the semiconductor behaves as a dielectric, losing its ability to store or conduct charge and effectively transforming into an insulator. The overall measured capacitance now comprises C_{ox} and the depletion layer capacitance, C_d, connected in series. Consequently, this arrangement leads to a reduction in the measured capacitance within the depletion region. As the gate voltage increases, the depletion region shifts away from the gate, effectively increasing the effective dielectric thickness between the gate and the substrate, which in turn diminishes the capacitance.

3.4.4 Inversion Region

As the applied voltage surpasses a threshold, a transition occurs where dynamic carrier generation and recombination shift toward net carrier generation. The applied positive voltage induces the formation of electron-hole pairs and attracts electrons, which are the minority carriers, toward the metal. Due to the insulating properties of the oxide, these minority carriers accumulate at the interface between the substrate and the oxide, forming a layer of minority carriers called the inversion layer, due to the reversed carrier polarity.

Beyond a certain positive gate voltage, the majority of available minority carriers reside within the inversion layer, and additional increases in gate voltage no longer lead to further semiconductor depletion. In essence, the depletion region reaches its maximum depth. Once the depletion region achieves its maximum depth, the high-frequency capacitance, shown as C_{HF} in Fig. 3.7, then consists of the series combination of C_{ox} and the maximum depletion capacitance, $C_{d,(max)}$, resulting in minimum capacitance. Consequently, the slope of the C-V curve becomes nearly flat. The inversion-region capacitance in the maximum depletion region (weak inversion) varies with measurement frequency, leading to varying C-V curve appearances. Differences are more pronounced at lower frequencies and diminish at higher frequencies.

3.4.5 C-V-Method Extracted Intrinsic Parameters

Within the depletion region, the gradient of the $1/C_d^2$ plot allows the calculation of N_a, the p-type non-ionized acceptor concentration. The intercept allows the built-in potential, V_{bi}, to be estimated.

3.4.5.1   Barrier Height

Consequently, the C-V method's barrier height (Φ) is then [32–34]:

$$\Phi = V_{bi} + \frac{kT}{q}\left(1 + \ln\frac{N_V}{N_a}\right).$$

$$(3.43)$$

where $N_V \approx 1.04 \times 10_{19}/\text{cm}^3$ is 300 K p-Si valence band density of states.

3.4.5.2   Interface State Density

The density of interface states D_{it}, also sometimes labeled as N_{ss}, is given by [31]

$$D_{it} = \frac{2}{qA}\frac{(G_a/\omega)_{\text{max.}}}{\left[((G_a/\omega)_{\text{max.}}C_{ox})^2 + (1 - C_a/C_{ox})^2\right]},$$

$$(3.44)$$

in terms of diode area, A, the measurement angular frequency, ω, and the corrected capacitance, the oxide capacitance, and the peak conductance. Figure 2.10 showed the typical peaking in the conductance that indicated the presence of interface states.

3.4.5.3   Interface State Distribution

Equation 2.25 provides a route to estimate the interface state density distribution over applied bias for a given p-type Schottky diode. The energy of interface states, E_{it}, (sometimes written as E_{ss}) is measured relative to the top of the valence band, is

$$E_{it} - E_V = q\Phi - qV$$

$$(3.45)$$

The peak of the (G_{ss}/ω) versus ω plot occurs at $(\omega\tau) \approx 1.98$, where τ is the average lifetime of the interface charges. The plots of E_{it} with applied bias are equivalent to $(E_{it} - E_V)$. Some recent studies suggest that the interface state density shows a sensitivity to temperature. Also, E_{it} has been shown by many recent studies to deviate substantially from linearity for larger biases. This evidence further supports that Φ also depends on bias in addition to the temperature.

3.4.6   Other C-V Measurement Approaches

Since the first CV approach was proposed, it has remained fundamentally unchanged. Numerous other CV methods are also suggested in the literature. Modifications of the method have mainly aimed at fine adjustments for specific measurement applications. For instance, Liu et al. [35] recently introduced what they named the two-step six-configuration parameter extraction method (TSPEM) for SBD parameters aimed at mitigating the impact of parasitic measurement parameters on the intrinsic capacitance-voltage (C-V) characteristics. Parasitic impedances are pronounced at high frequencies. This is useful in the creation of an accurate diode simulation model of SBD behavior. The TSPEM method takes two steps i.e., de-embedding and parameter extraction, employing six auxiliary configurations. Reportedly, the method is more accurate than traditional C-V methods.

Another example of a modernization of the C-V method is from Matacena et al. [36], who introduced a technique that utilizes the analysis of forward bias capacitance to characterize interfaces in Schottky structures. They focused on the specific case of a graphene/silicon solar cell. They observed multiple peaks in the experimental capacitance-voltage curve that they linked to non-uniform interface properties, such as localized defects. They developed an experimentally calibrated numerical model of the solar cell that accounted for non-uniform interfaces and provided knowledge of the distribution of defects.

3.4.7 Photoconductance and Photocapacitance

The three most commonly used semiconductor junction diode detectors are of the following types: p-n, MS, and metal-insulator (MIS). There are other types of photodetectors, such as the avalanche photodiode and various photoconductive sensors, but our main focus here is on the first three.

1. **p-n junction photodiode**: This device is composed of a shallow diffused p-region and an n-type active layer. When photons with energy exceeding the semiconductor's bandgap strike the neutral regions on both sides of the junction, they create electron-hole pairs (e-h pairs). These carriers, generated within a certain distance from the junction, move through diffusion to the space charge region, where they are separated by an electric field, ultimately producing a photocurrent. Figure 3.9 shows the energy band diagram of the p-n junction photodiode.
2. **MS junction photodiode**: A Schottky barrier photodiode (SBPD) is a majority carrier device that is similar to the p-n junction. A potential barrier separates the optically generated electron-hole (e-h) pairs. Carriers can be generated in both the neutral semiconductor region and the depletion region at the junction. However, compared to p-n junction photodiodes, SBPDs have several advantages, including simpler fabrication (metal barrier deposition on n or p semiconductor), absence of high-temperature diffusion processes, and high response speed. However, they tend to have relatively high dark currents [37]. Figure 3.10 shows the energy band diagram of the p-n junction photodiode.
3. **MIS junction photodiode**: The MIS photodiode features an insulating layer that separates a metal gate from a semiconductor surface. The critical factor in this photodiode design is maintaining an insulation layer thickness of around 10 nm

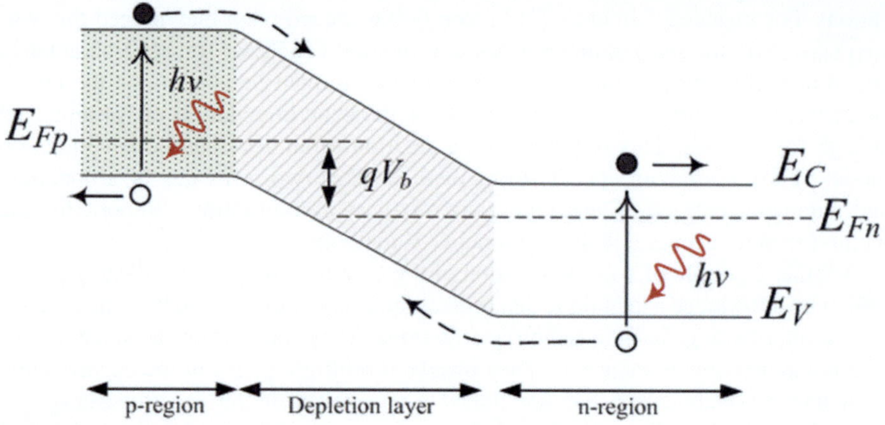

Fig. 3.9 Energy band diagram of the p-n junction photodiode

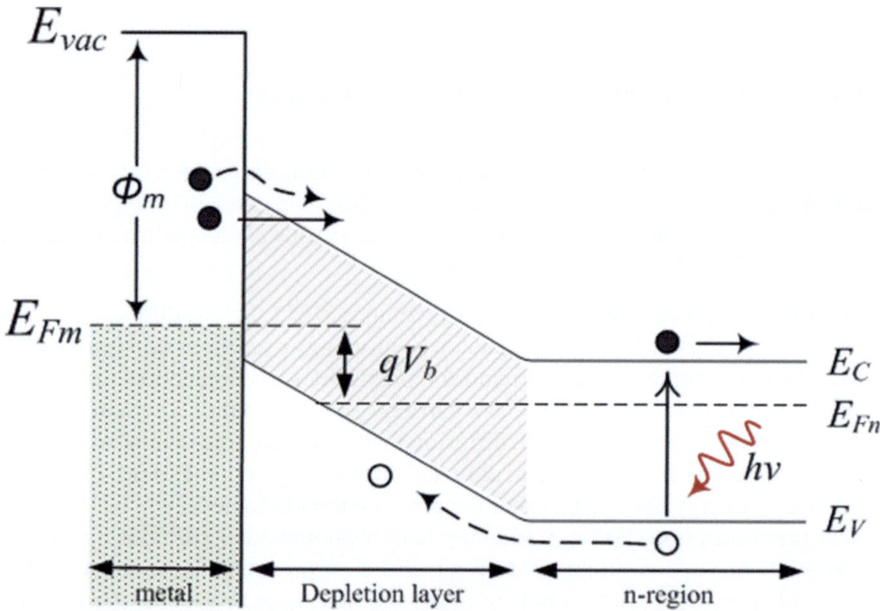

Fig. 3.10 Energy band diagram of the MS junction photodiode

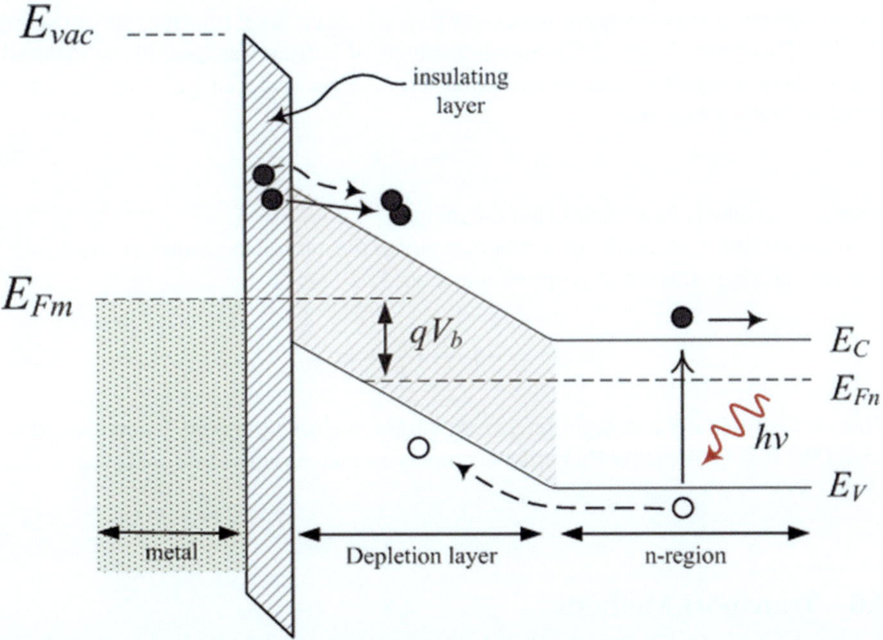

Fig. 3.11 Energy band diagram of the MIS junction photodiode

or more to prevent quantum electron tunneling. By applying a negative voltage to the metal electrode, electrons are pushed away from the insulator-semiconductor interface, leading to the creation of a depletion region. The well capacity, representing the total charge that the photogate can gather, relies on gate polarization, insulator thickness, electrode surface, and the semiconductor's background doping. Figure 3.11 shows the energy band diagram of the p-n junction photodiode.

3.5 Activation Energy Methods

Activation energy, E_a, in the semiconductor context, is the energy required to move an electron from the valence band to the conduction band. Before excitation, the electrons are bound to atoms in the valence band. After excitation, they jump to the conduction band and can participate as free-charge carriers. The activation energy is a crucial parameter in determining the electrical conductivity of a semiconductor. In intrinsic (pure) semiconductors, the activation energy is related to the bandgap energy (E_g).

In doped semiconductors, the activation energy can also refer to the energy required to move an electron from the valence band to the energy level of the accep-

tor or a donor. It then becomes necessary to apply deep-level transient spectroscopy
(DLTS) This gives rise to different mechanisms of current transport in the material.
The Arrhenius equation can be written in terms of the material's electrical conduc-
tivity, σ, and a factor σ_0:

$$\sigma = \sigma_0 e^{-\frac{E_a}{kT}}, \tag{3.46}$$

where the symbols have their usual meanings.

To determine E_a from the experiment, a plot of the natural logarithm of σ measured
at different temperatures, T, is plotted against $1/T$, i.e.,

$$\ln(\sigma) = -\left(\frac{E_a}{k}\right)\frac{1}{T} + \ln(\sigma_0). \tag{3.47}$$

This plot is typically a straight line, from whose negative gradient E_a can be calculated [38]. It is evident that Richardson's plot is, in fact, an activation energy approach
[39].

3.6 Transport Methods

The TE model is considered to most successfully explain the current transport mechanisms in many materials. However, as is now evident, it is subjected to many con-
straining assumptions. These assumptions generalize a given device to near ideality.
The TE model suffers from irregularities in the barrier and other deviations such
as considerable series resistance. In the event of such deviations, it is difficult to
completely establish the transport mechanism in the device [40].

In semiconductors, there are several types of mechanisms that define the current
transport within a given material. Each mechanism occurs under specific conditions:

1. Drift: Drift current is the movement of charge carriers (electrons or holes) in
   response to an electric field. It occurs when a voltage is applied across the semiconductor, causing carriers to move in the direction of the field. Drift current is
   the dominant mechanism in electronic devices like diodes and transistors.
2. Diffusion current: Diffusion current is driven by the concentration gradient of
   charge carriers within the semiconductor. It occurs due to the tendency of carriers
   to move from regions of higher concentration to regions of lower concentration.
   Diffusion current is a key factor in bipolar junction transistors (BJTs) and can be
   important in certain types of diodes.
3. Thermal generation and Recombination: Intrinsic carriers (electrons and holes)
   can be thermally generated within the semiconductor lattice due to the energy
   provided by temperature. These carriers can participate in the current flow. Additionally, carriers can recombine, leading to a loss of carriers and reduced current.
   These processes are significant in the operation of solar cells and other devices
   sensitive to temperature.

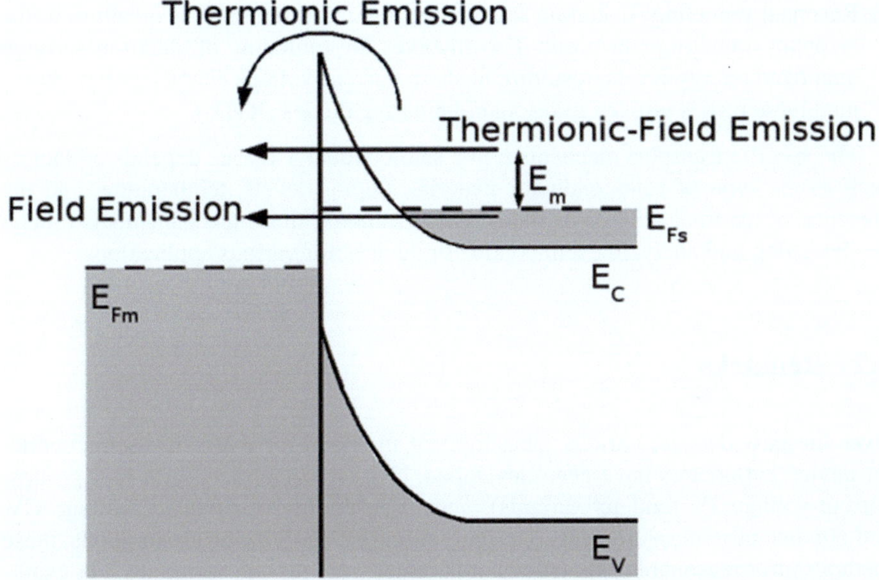

Fig. 3.12 Common carrier transport mechanisms in the SBD in forward bias [41]

4. Trap-assisted tunneling: In the presence of traps or energy states within the bandgap of the semiconductor, trap-assisted tunneling can occur. Charge carriers can tunnel between traps and the conduction or valence band, contributing to the current. This is important in tunnel diodes and certain types of memory devices.
5. Field emission: Field emission involves the emission of electrons from a semiconductor surface into a vacuum or a different material when a strong electric field is applied. This phenomenon is used in field-effect transistors (FETs) and electron emission devices like cold cathodes. Figure 3.12 shows the process of field emission.
6. Thermionic emission: Thermionic emission occurs when electrons overcome the energy barrier at the semiconductor-metal interface and enter the metal. This is typical in metal-semiconductor contacts and is used in thermionic emission diodes. Figure 3.12 shows the process of thermionic emission.
7. Avalanche and Zener breakdown: In reverse bias conditions, if the electric field across a semiconductor becomes strong enough, it can lead to avalanche or Zener breakdown. In these processes, electrons gain enough energy to create electron-hole pairs through impact ionization, leading to a sudden increase in current. Avalanche breakdown is used in avalanche photodiodes, while Zener breakdown is exploited in Zener diodes for voltage regulation.

8. Resonant tunneling: In certain semiconductor structures, such as quantum wells, resonant tunneling can occur. This involves the tunneling of electrons through quantized energy levels, resulting in sharp current-voltage characteristics. Resonant tunneling is utilized in resonant tunneling diodes (RTDs).

The specific transport mechanism in a semiconductor device depends on factors such as the type of semiconductor material, doping levels, temperature, and the presence of external voltage or fields. Understanding these mechanisms is crucial for designing and analyzing semiconductor devices for various applications.

3.7 Remarks

Over the past decade, various measurement methods for extracting semiconductor device parameters have been developed, broadly categorized into four groups: current-voltage IV (and its variants), activation energy, capacitance-voltage CV, and photoelectric measurements. Despite relying on simplifying assumptions, these methods offer reasonable estimations of parameter ranges and variations. For example, the C-V method is commonly applied in reverse bias using high-frequency excitation signals, with an assumption that the diode won't exhibit a low-voltage resonance peak caused by interfacial charges. Meanwhile, the I-V method necessitates the consideration of the diode's forward resistance and presents three categories of extraction methods: simplistic models based on transport equations, physical analytical models, and numerical models relying on equations like Poisson's and drift-diffusion. Many existing approaches impose constraints on parameter values and resistance values, limiting their applicability. This chapter focuses on newer methods, emphasizing those addressing series resistance (R_s) and interface trap density due to their significant impact on MS device performance.

References

1. H. Wang, X. Chen, G. Xu, K. Huang, A novel physical parameter extraction approach for Schottky diodes. Chin. Phys. B **24**(7), 077305 (2015). https://doi.org/10.1088/1674-1056/24/7/077305
2. S.J. Moloi, M. McPherson, Capacitance-voltage behaviour of Schottky diodes fabricated on p-type silicon for radiation-hard detectors. Radiat. Phys. Chem. (2013). https://doi.org/10.1016/j.radphyschem.2012.12.002 (*Article in press*)
3. A.I. Prokopyev, S.A. Mesheryakov, Restrictions of forward I-V methods for determination of Schottky diode parameters. Measurement **33**(2), 135–144
4. S.M. Sze, K.K. Ng, Physics of Semiconductor Devices, 3rd edn. (Wiley, 2007)
5. I. Dökme, P. Durmus, S. Altindal, Effects of γ-ray irradiation on the C-V and G/ω-V characteristics of Al/SiO$_2$/p-Si (MIS) structures. Nucl. Instrum. Methods Phys. Res. Sect. B: Beam Inter. Mater. Atoms **266**(5), 791–796 (2008)
6. S. Karatas, A. Türüt, Electrical properties of Sn/p-SI(MS) Schottky barrier diodes to be exposed to ^{60}Co γ-ray source. Nucl. Instr. Meth. A **566**, 584–589 (2006)

7. A.I. Prokopyev, S.A. Mesheryakov, Fast extraction of static parameters of Schottky diodes from forward I-V characteristic. Measurement **37**(2), 149–155 (2005)
8. H. Norde, A modified I-V plot for Schottky diodes with high series resistance. J. Appl. Phys. **50**(7), 5052–5053 (1979)
9. C.-D. Lien, F.C.T. So, M.A. Nicolet, An improved forward I-V method for nonideal Schottky diodes with high series resistance. IEEE Trans. Electron Devices **ED-31**(10), 502–1503 (1984)
10. J.H. Werner, Schottky barrier and pn-junction I/V plots-small signal evaluation. Appl. Phys. A **47**(3), 291–300 (1988)
11. H. Norde, A modified forward IV plot for Schottky diodes with high series resistance. J. Appl. Phys. **50**(7), 5052–5053 (1979)
12. S.K. Cheung, N.W. Cheung, Extraction of Schottky diode parameters from forward current-voltage characteristics. Appl. Phys. Lett. **49**(2), 85–87 (1986)
13. A.B. McLean, Limitations to the Norde IV plot. Semicond. Sci. Technol. **1**(3), 177 (1986). https://doi.org/10.1088/0268-1242/1/3/003
14. K. Sato, Y. Yasumura, Study of forward I-V plot for Schottky diodes with high series resistance. J. Appl. Phys. **58**(9), 3655–3657 (1985). https://doi.org/10.1063/1.335750
15. K.E. Bohlin, Generalized Norde plot including determination of the ideality factor. J. Appl. Phys. **60**(3), 1223–1224 (1986). https://doi.org/10.1063/1.337372
16. D.E. Yıldız, A. Kocyigit, M. Yıldırım, Comparison of Al/TiO2/p-Si and Al/ZnO/p-Si photodetectors. Opt. Mater. **145**, 114371 (2023). https://doi.org/10.1016/j.optmat.2023.114371
17. R.O. Ocaya, F. Yakuphanoğlu, Ocaya-Yakuphanoğlu method for series resistance extraction and compensation of Schottky diode I-V characteristics. Measurement **186**, 110105 (2021). https://doi.org/10.1016/j.measurement.2021.110105
18. A. Demirci, H.G. Çetinkaya, P. Durmuş, S. Demirezen, Ş. Altındal, Optoelectronic characterization of Bi-doped ZnO nanocomposites for Schottky interlayer applications. Phys. B: Condens. Matter 415338 (2023). https://doi.org/10.1016/j.physb.2023.415338
19. V. Mikhelashvili, R. Padmanabhan, G. Eisenstein, Simplified parameter extraction method for single and back-to-back Schottky diodes fabricated on silicon-on-insulator substrates. J. Appl. Phys. **122**(3) (2017). https://doi.org/10.1063/1.4994176
20. H. Kasani, Fabrication of carbon nanoparticle/polymer nanocomposite-based thermometer: communication. Diyala J. Eng. Sci. 606–621 (2015)
21. V. Mikhelashvili, G. Eisenstein, V. Garber, S. Fainleib, G. Bahir, D. Ritter, M. Orenstein, A. Peer, On the extraction of linear and nonlinear physical parameters in nonideal diodes. J. Appl. Phys. **85**(9), 6873–6883 (1999). https://doi.org/10.1063/1.370206
22. H. Durmuş, Ü. Atav, Extraction of voltage-dependent series resistance from IV characteristics of Schottky diodes. Appl. Phys. Lett. **99**(9), 093505 (2011). American Institute of Physics. https://doi.org/10.1063/1.3633116
23. R. Nouchi, Extraction of the Schottky parameters in metal-semiconductor-metal diodes from a single current-voltage measurement. J. Appl. Phys. **116**(18) (2014). https://doi.org/10.1063/1.4901467
24. O. Donagh, R. Duane, T. Campagno, L. Lewis, N. Cordero, P. Maaskant, F. Waldron, B. Corbett, Thermal stability of SiC Schottky diode anode and cathode metalisations after 1000 h at 350°C. Microelectr. Reliab. **51**, 904–908 (2011). https://doi.org/10.1016/j.microrel.2010.12.008
25. R.O. Ocaya, A. Al-Ghamdi, F. El-Tantawy, W.A. Farooq, F. Yakuphanoglu, Thermal sensor based zinc oxide diode for low-temperature applications. J. Alloys Compounds **674**, 277–288 (2016). https://doi.org/10.1016/j.jallcom.2016.02.267
26. E. H. Rhoderick, R.H. Williams, *Metal-Semiconductor Contacts* (Clarendon Press, 1988)
27. W.A. Hill, C.C. Coleman, A single-frequency approximation for interface-state density determination. Solid-State Electr. **23**(9), 987–993 (1980). https://doi.org/10.1016/0038-1101(80)90064-7
28. İ. Døkme, Ş. Altindal, İ. Uslu, The effects of temperature, radiation, and illumination on current-voltage characteristics of Au/PVA (Co, Zn-doped)/n-Si Schottky diodes. J. Appl. Polymer Sci. **125**(2), 1185–1192 (2012). https://doi.org/10.1002/app.36327

29. İ. Døkme, Ş. Altındal, T. Tunç, İ. Uslu, Temperature dependent electrical and dielectric properties of Au/polyvinyl alcohol (Ni, Zn-doped)/n-Si Schottky diodes. Microelectr. Reliab. **50**(1), 39–44 (2010). https://doi.org/10.1016/j.microrel.2009.09.005

30. R.O. Ocaya, Y. Orman, A.G. Al-Sehemi, A. Dere, A.A. Al-Ghamdi, F. Yakuphanoğlu, Bias and illumination-dependent room temperature negative differential conductance in Ni-doped ZnO/p-Si Schottky photodiodes for quantum optics applications. Heliyon **9**(5) (2023). https://doi.org/10.1016/j.heliyon.2023.e16269

31. E.H. Nicollian, J.R. Brews, *MOS (Metal Oxide Semiconductor) Physics and Technology* (Wiley, New York, 1982)

32. A.N. Donald, *Semi-Conductor Physics & Devices* (Tata McGraw Hill Education Private Limited, 2006)

33. B. Gunduz, A.A. Al-Ghamdi, A.A. Hendi, Z.H. Gafer, S. El-Gazzar, F. El-Tantawy, F. Yakuphanoğlu, New Schottky diode based entirely on nickel aluminate spinel/p-silicon using the sol-gel spin coating approach. Superlattices Microstruct. **64**, 167–177 (2013). https://doi.org/10.1016/j.spmi.2013.09.022

34. M. Saremi, M. Saremi, H. Niazi, A.Y. Goharrizi, Modeling of lightly doped drain and source graphene nanoribbon field effect transistors. Superlattices Microstruct. **60**, 67–72 (2013). https://doi.org/10.1016/j.spmi.2013.04.013

35. X. Liu, Y. Zhang, H. Wang, L. Qi, B. Wang, J. Zhou, W. Ding, Z. Jin, F. Xiao, TSPEM parameter extraction method and its applications in the modeling of planar Schottky diode in THz band. Electronics **10**(13), 1540 (2021). https://doi.org/10.3390/electronics10131540

36. I. Matacena, P. Guerriero, L. Lancellotti, E. Bobeico, N. Lisi, R. Chierchia, P.D. Veneri, S. Daliento, Forward bias capacitance investigation as a powerful tool to monitor graphene/silicon interfaces. Solar Energy **226**, 1–8 (2021). https://doi.org/10.1016/j.solener.2021.08.016

37. Z. Bielecki, K. Achtenberg, M.E. Kopytko, J.A. Mikolajczyk, J. Wojtas, A.W. Rogalski, Review of photodetectors characterization methods. Bull. Polish Acad. Sci. Tech. Sci. **70**(2) (2022). https://doi.org/10.24425/bpasts.2022.140534

38. I. Gumus, S. Aydogan, Thermal sensing capability of metal/composite-semiconductor framework device with the low barrier double Gaussian over wide temperature range. Sens. Actuators A: Phys. **332**, 113–117 (2021). https://doi.org/10.1016/j.sna.2021.113117

39. S. Kim, T.H. Seo, M.J. Kim, K.M. Song, E.-K. Suh, H. Kim, Graphene-GaN Schottky diodes. Nano Res. **8**, 1327–1338 (2015). https://doi.org/10.1063/5.0064036

40. J.W. Kleppinger, S.K. Chaudhuri, O.F. Karadavut, K.C. Mandal, Role of deep levels and barrier height lowering in current-flow mechanism in 150 μm thick epitaxial n-type 4H-SiC Schottky barrier radiation detectors. Appl. Phys. Lett. **119**(6), 202106 (2021). https://doi.org/10.1063/5.0064036

41. K. Mensah-Darkwa, R. Ocaya, A. Dere, A.G. Al-Sehemi, A.A. Al-Ghamdi, M. Soylu, R.K. Gupta, F. Yakuphanoglu, Dye based photodiodes for solar energy applications. Appl. Phys. A **123**, 1–13 (2017). https://doi.org/10.1007/s00339-017-1221-x

Chapter 4
Transient Parameter Extraction Methods

Abstract Semiconductor traps or impurity states, interface states, or defects, play a crucial role in the performance and reliability of semiconductor devices. Understanding semiconductor traps is essential for optimizing device performance, enhancing device reliability, and improving the efficiency and functionality of new materials, particularly in optoelectronic devices. This chapter explores the significance of characterizing these traps, which are introduced into the crystal lattice through intentional or unintentional doping. It is seen that the method of determining traps has remained unchanged over the last decade, although there are important studies that require specific mention in this chapter. Furthermore, it becomes evident in this chapter that the ideal characterization technique should be sensitive, rapid, and analytically straightforward, capable of differentiating between majority- and minority-carrier traps. Spectroscopic capabilities for distinguishing signals from different traps, along with reproducible plotting, are crucial. Additionally, the method should assess traps at various depths, encompassing both radiative and nonradiative centers. Luminescence spectroscopy is effective for investigating shallow centers, while more sophisticated techniques are required for deeper nonradiative recombination centers, which often dominate recombination kinetics in many devices.

4.1 Photoconductance and Photo Capacitance

The study of impurity-based carrier trapping levels remains central in research into semiconductors. Today, traps are referred to in many terms, such as impurity states, interface states, defects, and so on. Defects in the crystal lattice introduce localized states within the band gap, either unintentionally or intentionally through dopants. As the scale of new devices becomes ever smaller and new potential semiconductors are discovered almost daily, it is crucial to characterize these traps.

Apart from being fundamental research into extending our grasp of their behaviors, the study of semiconductor traps is important for several reasons:

1. Device performance—traps are generally undesirable since they significantly affect device performance. Understanding them may lead to mitigation techniques for optimizing device performance. Also, devices with abundant traps perform less reliably and fail more rapidly due to degradation over time.
2. New materials—traps are essentially energy loss centers in semiconductor devices. In new materials especially, the correct characterization of the distribution and density of these trap states is important for improved device efficiency and functionality. For instance, in optoelectronic devices i.e. photodetectors and photovoltaics, traps remove photogenerated carriers from the external circuit. This manifests as a lowered signal current level, or open-circuit voltage.

The ideal technique for characterizing traps in a semiconductor should be sensitive, rapid, and analytically straightforward. It should distinguish between majority- and minority-carrier traps and provide details about their concentrations, energy levels, and capture rates. Spectroscopic capabilities, which allow distinguishing signals from different traps and reproducible plotting, are essential. Additionally, the method should measure traps at various depths, including radiative and nonradiative centers. Luminescence is effective for shallow centers while deeper nonradiative centers are not yet fully understood and require more sophisticated techniques. They are considered to be more important in many devices since they tend to dominate the recombination kinetics.

Figure 4.1 shows five recombinations that involve defect centers in a simplified, flat-band representation.

Semiconductor recombination mechanisms can be categorized into five types: radiative recombination (also called band-to-band recombination), Shockley-Read-Hall recombination, Auger recombination, surface recombination, and Langevin recombination. Radiative recombination is particularly prominent in indirect band gap semiconductors [1]. Band-to-band, also called bimolecular, recombination, is common in direct band gap materials and involves low photon momentum, this type of recombination is significant in the direct band gap semiconductors.

Surface recombination is characterized by a surface recombination velocity that relies on the surface defect densities. This phenomenon plays a central role in solar cells, where the collection and extraction of free charge carriers occur at the surface. To mitigate surface recombination in solar cells, a thin transparent window layer with a wider band gap is employed, and passivation processes are applied to minimize this type of recombination. In recent years, there has been an increased focus on surface passivation techniques to minimize this type of recombination, especially in 2D devices [3]. On the other hand, the Langevin recombination mechanism is observed in organic semiconductors due to low-mobility charge carriers of opposite polarity interfering with each other.

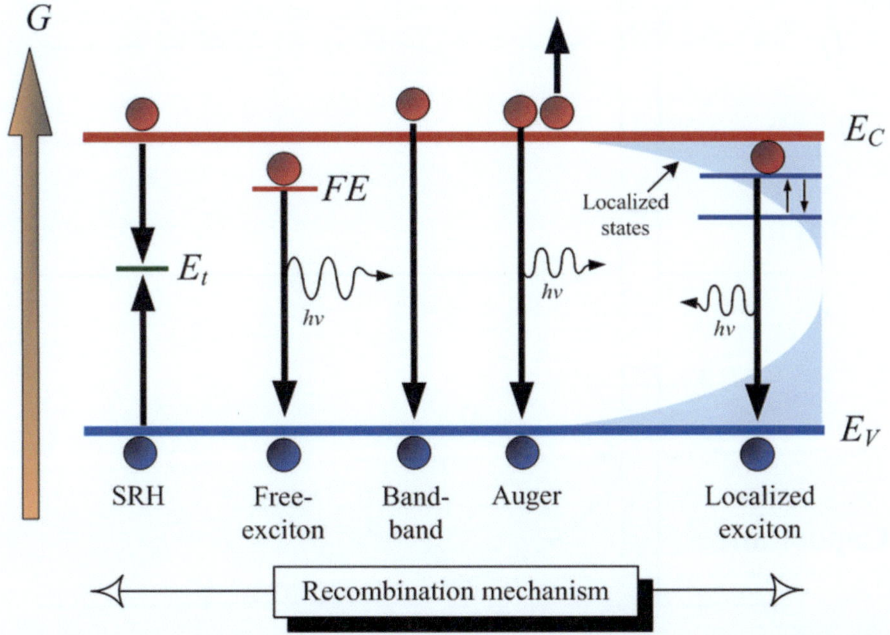

Fig. 4.1 Depiction of five different recombination mechanisms, some of which are non-radiative. The electrons are shown in red and the holes in blue. Adapted from Huang et al. [2]

4.1.1 Characterizing Deep-Energy Levels

Photo capacitance spectroscopy entails measuring AC capacitance while varying the wavelength of illumination. When the photon energy matches the activation energy of a deep center, carriers are released from this center to the corresponding band edge. These carriers then redistribute, affecting the sample's polarization and resulting in a capacitance peak.

Deep-level transient spectroscopy (DLTS) is employed to study charge carrier traps in the depletion region of Schottky and p-n junction diodes. The technique was first used by Richard Williams in a 1966 study of deep-traps in GaAs using gated voltage-induced transients [4]. It is a well-established technique that continues to find application in characterizing new semiconductor materials [5–7]. Figure 4.2 shows the reverse biasing voltage-pulse technique employed by Williams and others.

The DLTS setup permits measuring the time constant of several transients pertaining to a given trap level at different temperatures. This leads to a plot of τ versus $1000/T$ from which the activation energy of the trap level can be found. The initial height of the capacitance transient is proportional to the trap density, N_o.

Today, DLTS is well-adapted to light pulse stimulation and is useful for its speed of extracting information about the trap levels. When photon energy matches the activation energy of a deep center, it releases carriers from that center to the adjacent

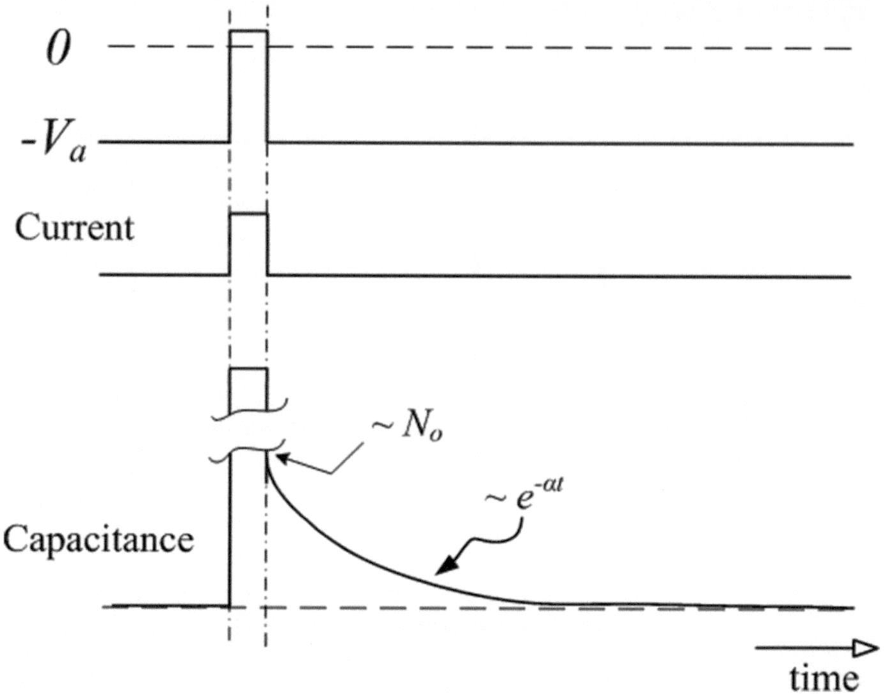

Fig. 4.2 Depiction of a capacitance transient response to a reverse-bias voltage pulse, V_a. This was the first approach to transient spectroscopic measurements of deep-levels [4]. This diagram shows the specific case of a p-type material i.e. the majority-carriers are holes. The time constant of the transient, i.e. $\tau = 1/\alpha$, is a function of temperature. The concentration of the given trap is related to the initial capacitance height

band edge. These carriers then redistribute, either as free carriers or become trapped in another location due to bias. This redistribution alters the sample's polarization, resulting in the observation of a capacitance peak [8, 9]. It determines key defect parameters and their concentrations in the depletion region of electronic devices, typically Schottky diodes or p-n junctions, by briefly altering the reverse polarization voltage. This disturbance allows carriers from the semiconductor to recharge the defects, leading to subsequent emission of trapped carriers and observable capacitance transients. Cycling through voltage pulses and charge state recovery enables analysis of the defect recharging process.

Semiconductor materials respond to pulsed illuminations through transients that yield useful information about their photonic behaviors. Transients are not instantaneous evolutions of responses owing to the underlying mechanisms of conduction in the materials. These transients, which are typically observed in the current, voltage, or capacitance, provide valuable information about the charge carrier dynamics, recombination processes, and energy levels within the material. In a sense, transient measurements are a subset of techniques called electrical spectroscopy that are a specific case of the activation energy method discussed in Chap. 3.

4.1.2 Equilibration

Before the sample is illuminated, an equilibration procedure is followed. This is to allow the sample to attain its thermal equilibrium state with no excess charge carriers in the material [4]. Then, the illumination is applied, generating e-h pairs and a temporary increase in charge density. The transient measurements are then fitted to a specifically chosen model that describes the rate of decay of excess carriers, allowing the extraction of the fitting parameters, such as the carrier lifetime, τ. Exponential decay or biexponential decay models are typically used since they lead rapidly to τ, the average time it takes for the excess charge carriers to recombine or decay to their equilibrium state.

The specifics of a given photo capacitance measurement and extraction procedure depend on the experimental setup, the studied semiconductor, and the characteristics of the photo capacitance response. Advanced techniques such as time-resolved photo capacitance spectroscopy (TRPCS) may be employed to gain more detailed information about carrier dynamics.

4.2 Measuring Photocurrent and Photo Capacitance

A thin film of the semiconductor material is first prepared and then exposed to a source of illumination of energy greater than the material band gap. The source is pulsed with an envelope of low-repetition rate but modulated at a higher frequency. For instance, a pulse train of 10 Hz whose ON-times are modulated with 10–100 kHz. TPC measurements are vital for charge carrier dynamics in photovoltaic materials tied to overall solar cell efficiency.

The light source can be a laser or a pulsed light-emitting diode (LED) of minimum energy, $h\nu$, (maximum wavelength) to generate electron-hole pairs in the semiconductor material. The effect of illumination on capacitance can be expressed in terms of the photocurrent (I):

$$C_{\text{diff}} = \frac{\tau}{V_T} I \tag{4.1}$$

where τ is the mean lifetime of the minority carriers [10, 11].

As discussed in earlier chapters, capacitance-based measurements are done through the $C - V$ measurements that must be compensated for the effects of series resistance and the true depletion layer capacitance in the device. These parameters are frequency-sensitive [12]. The adjusted capacitance (C_a) and conductance (G_a) in the presence of series resistance and actual depletion layer are usually calculated to factor the frequency sensitivities of junction capacitance and series resistance in the real part of the impedance [13–15].

The actual measurement setup emulates a capacitor i.e. it is fitted with electrodes, but windowed to allow incidence of the illumination. Upon illumination, the generated electron-hole pairs within the material contribute to the charge stored in the capacitor, leading to an observable change in capacitance over time, hence the term transient photo capacitance (TPC). The transient response of the capacitance is useful for deducing the charge carrier dynamics, their lifetimes, interface-state density, mobility, energy levels, and general insight into the active recombination processes.

Transient C-V and DLTS methods rely on the thermal excitation of holes or electrons from defect levels within the bandgap. The thermal emission rates of holes and electrons, denoted as e_p and e_n, from the defect levels can be expressed as follows:

$$e_p = \sigma_p v_p N_V e^{\left(\frac{E_V - E_T}{kT}\right)}, \tag{4.2}$$

$$e_n = \sigma_n v_n N_C e^{\left(\frac{E_T - E_C}{kT}\right)}, \tag{4.3}$$

where σ_p and σ_n represent the thermal capture cross-sections of holes and electrons at the defect level, v_p and v_n are their thermal velocities, N_V and N_C are the effective density of states in the conduction and valence bands with band edges E_V and E_C, and E_T is the effective energy level of the trap.

The capture cross-section, σ, of a trap energy level signifies the likelihood that a free charge carrier (electron or hole) will be trapped when encountering the energy level within a semiconductor's bandgap. This parameter is crucial for understanding charge carrier behavior and impacts various electrical properties in semiconductor devices. It quantifies the probability of capture and is often expressed as a function of energy and temperature [16].

4.2.1 Reverse-Bias Photoresponses

Figure 4.3 illustrates a possible scenario during the operation of the traps under external illumination. The diagram shows a deep and a shallow level at E_d and E_s, respectively. Under continuous illumination, the process of electron generation occurs continuously in the order of generation in the valence band, transitioning to the conduction band through the shallow level, then trapping by the shallow level, and then possibly by the deep level.

In the conventional photo capacitance model, an underlying assumption is that all carriers residing at defect levels will eventually emit to the semiconductor's band edge. Consequently, this leads to a constant saturation photo capacitance, C_0. Under this assumption, C_0 remains unaffected by variations in the optical excitation power and is directly proportional to the defect concentration. However, Zhang et al. [17] showed that there tends to be a notable dependence of the saturation photo capac-

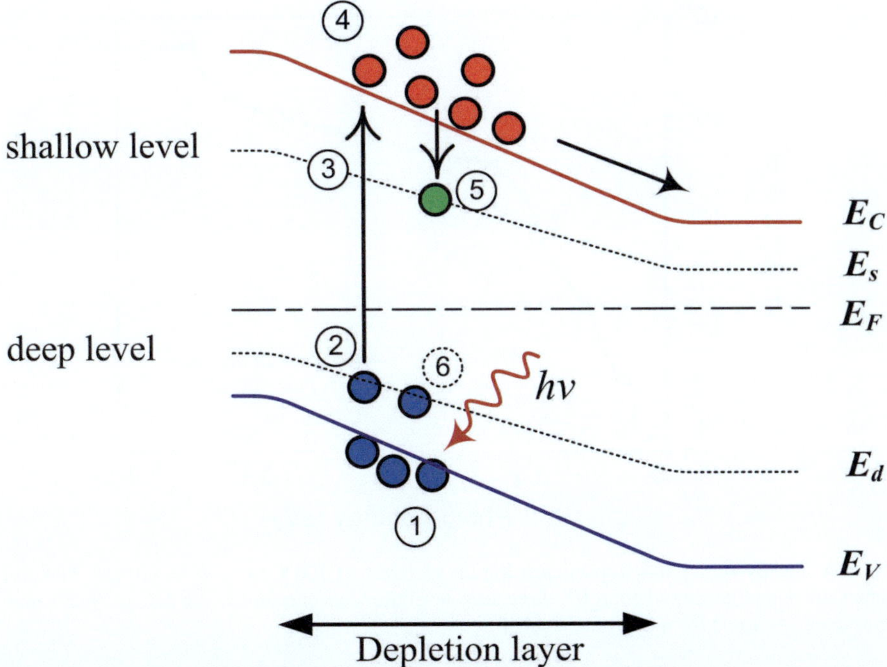

Fig. 4.3 A representation of shallow level (E_s) and deep level (E_d) traps at the junction of an n-type Schottky diode, on the semiconductor side. The energy band diagram is shown only with the traps illuminated for the sake of simplicity. The photo-generated electrons reach E_c through E_d and E_s. Then they are recaptured by E_s. Thus, the carrier trapping sequence is $1 \rightarrow 2 \rightarrow \cdots \rightarrow 6$. The sequence $5 \rightarrow 6$ (i.e. recapturing from shallow levels into the deep level E_d) is considered unlikely due to the low capture cross-section

itance on the optical excitation power and suggested a photo capacitance model to accommodate these observations.

In the conventional model, the transient photo capacitance is given by [18, 19]:

$$\Delta C(t) = \Delta C_{ss}[1 - e^{-e_n t}], \tag{4.4}$$

where ΔC_{ss} is the saturation photo capacitance.

The deep levels can recapture electrons, although this does not generally happen due to their low capture cross-section. Furthermore, the field associated with the Schottky barrier effectively drives most conduction band electrons out of the depletion layer region. However, a fraction of photoionized electrons is recaptured by the shallow level since it resides above the Fermi level, E_F, within the depletion region and remains unoccupied. Subsequently, these electrons at the shallow level recombine with holes located at the deep level. Figure 4.4 is a useful adapted steady-state photo capacitance for a semiconductor sample. Figure 4.5 shows the variations of

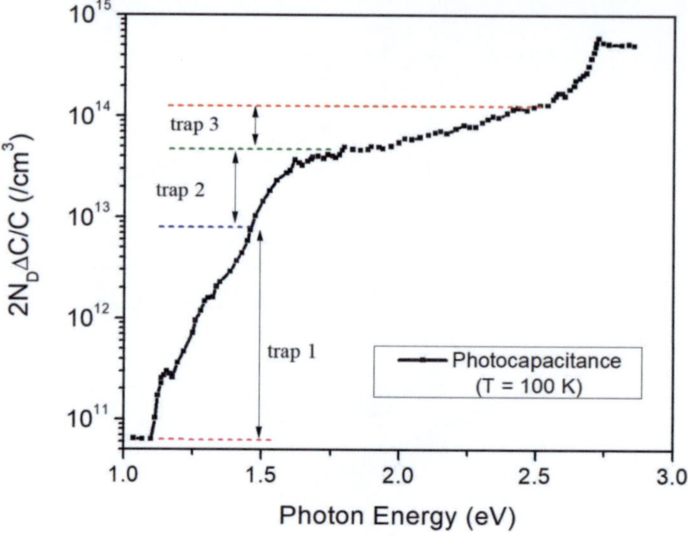

Fig. 4.4 Steady-state photo capacitance log-lin spectrum at 100 K for a ZnSe sample. The plot shows the optical threshold energies, three deep-level trap concentrations, and the near-band-edge emissions. Adapted from Hierro et al. [19]

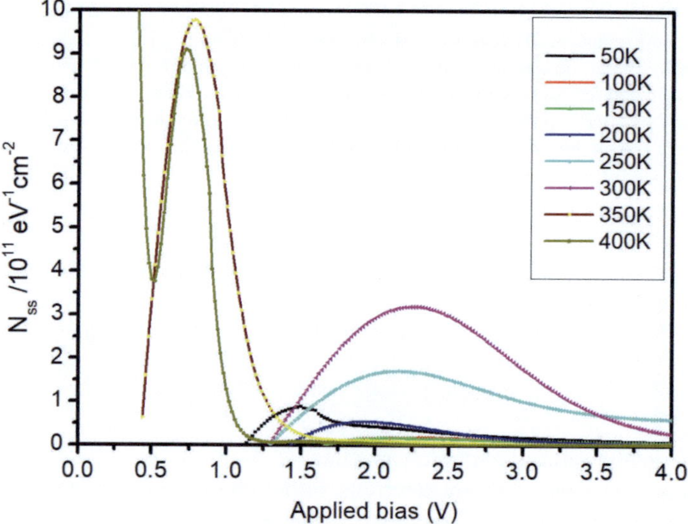

Fig. 4.5 An illustration of the density of interface states from an actual p-type Schottky diode [20]

the experimentally determined trap state densities for an organic on p-Si Schottky diode. The strong temperature dependence of the interface state density is evident in the plots.

There have been alternative approaches to the derivation of the transient photo capacitance by many workers. For the sake of brevity, we shall consider only the noteworthy contribution devised by Zhang et al. [17]. They began by assuming that the density of electrons at the deep level is n_d at time t, and the concentrations of the deep and shallow level centers are, respectively, n_t. Then, the density of electrons at the deep level is:

$$n_t = \alpha(N_T - n_t), \tag{4.5}$$

where α is a proportionality constant.

The rate equation governing electron emission and recapture at the deep level is formulated as:

$$\frac{dn_t}{dt} = -e_n n_t + r(N_T - n_t)\alpha(N_T - n_t), \tag{4.6}$$

where r represents the recombination coefficient for electrons at the shallow level recombining with holes at the deep level, and e_n ($=1/\tau$) is the rate of emission of electrons from the deep level, given by:

$$e_n = \phi\sigma, \tag{4.7}$$

where ϕ is the incident photon flux that is proportional to the incident illumination power P, and inversely to the cross-sectional photoactive area A of the capacitor, as shown in Fig. 4.6.

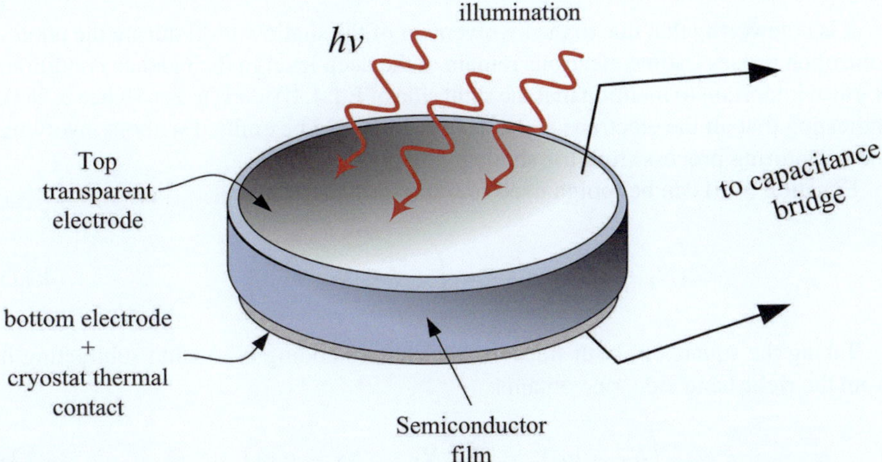

Fig. 4.6 Illustration of the transient photo capacitance measurement setup. The temperature of the sample is controlled using a cryostat in good thermal contact

In Eq. 4.6, the first term on the right-hand side signifies the reduction of electrons at the deep level due to photoionization, while the second term represents the incrementation of electrons due to recapturing from the shallow level.

As N_T remains constant, Eq. 4.6 can be rewritten as:

$$-\frac{d(N_T - n_t)}{dt} = e_n(N_T - n_t) - e_n N_T + r\alpha(N_T - n_t)^2.$$

(4.8)

Considering the initial condition at $t = 0$, $N_T - n_t = 0$, the solution for Eq. 4.6 is given by:

$$N_T - n_t(t) = a_1 \left(\frac{1 - a_2 e^{-a_3 t}}{1 + a_2 e^{-a_3 t}}\right) - a_4,$$

(4.9)

where

$$a_4 = \frac{e_n}{2r\alpha},$$
$$a_1 = \left(a_4^2 + 2a_4 N_T\right)^{1/2},$$
$$a_3 = 2r\alpha a_1,$$
$$a_2 = \frac{a_1 - a_4}{a_1 + a_4}.$$

Assuming that the density of electrons at the deep level reaches n_{t0} at equilibrium and setting $t = \infty$ in Eq. 4.9, we obtain:

$$N_T - n_{t0} = a_1 - a_4 = \left(\frac{e_n^2}{4r^2\alpha^2} + \frac{e_n N_T}{r\alpha}\right)^{1/2} - \frac{e_n}{2r\alpha}.$$

(4.10)

It is noteworthy that due to the involvement of the shallow level during the photoionization process, some electrons remain at the deep level in the balance condition. It's also important to mention that the right side of Eq. 4.10 tends to zero when $\alpha = 0$, indicating that all the electrons at the deep level would be emitted without involving the recapturing process from the shallow level.

Equation (4.6) can be reorganized into the form:

$$(N_T - n_{t0}) + \frac{e_n}{2r\alpha} = \left(\frac{e_n^2}{4r^2\alpha^2} + \frac{e_n N_T}{r\alpha}\right)^{1/2}.$$

(4.11)

Taking the squares of both sides of Eq. 4.11 and adding N_T to and subtracting it from the right-hand side, one obtains:

$$(N_T - n_{t0})^2 = \frac{e_n}{r\alpha}\left(N_T - (N_T - n_{t0})\right).$$

(4.12)

This equation can also be obtained by setting $dn_t/dt = 0$ in Eq. 4.6. Under weak optical excitation, $(N_T - n_{t0}) \ll N_T$ s.t.

$$N_T - n_{t0} \approx \left(\frac{e_n}{r\alpha} N_T\right)^{1/2}. \tag{4.13}$$

The total incremental density of charge in the depletion region is given by:

$$\Delta N = N_T - n_t - n'_t = (1 - \alpha)(N_T - n_t). \tag{4.14}$$

The change in capacitance induced by the increased depletion layer density is given by [19]:

$$\Delta C = \left(\frac{\Delta N}{2N_D}\right) C, \tag{4.15}$$

where C is the equilibrium capacitance without light excitation, and N_D is the donor concentration of the material.

Using Eqs. 4.9, 4.14, and 4.15, the fractional change in the transient photo capacitance at a given time instant is then:

$$\frac{\Delta C}{C} = \left(\frac{1-\alpha}{2N_D}\right) \left\{ a_1 \frac{1 - a_2 e^{-a_3 t}}{1 + a_2 e^{-a_3 t}} - a_4 \right\}. \tag{4.16}$$

Equations 4.13, 4.14 and 4.15 lead together allow the fractional transient *saturation* photo capacitance, $\Delta C_0/C$ to be expressed as:

$$\frac{\Delta C_0}{C} = \left(\frac{1-\alpha}{2N_D}\right) (N_T - n_{t0}) \tag{4.17}$$

$$\approx \left(\frac{1-\alpha}{2N_D}\right) \left(\frac{e_n N_T}{r\alpha}\right)^{1/2} \tag{4.18}$$

$$\therefore \frac{\Delta C_0}{C} \approx \left(\frac{1-\alpha}{2N_D}\right) \left(\frac{e_n N_T}{r\alpha A h \nu}\right)^{1/2} P^{1/2}. \tag{4.19}$$

In the last equation, P is the optical illumination power, which has photon energy $h\nu$, and A is the device area. The above equation predicts a linear relationship between the saturation photo capacitance and the square root of the optical power under weak excitation.

In order to compare with the conventional photo capacitance model, Zhang et al. [17] rewrote the transient photo capacitance, Eq. 4.16, as

$$\Delta C(t) = \Delta C_0 \left(1 - \frac{2a_1 a_2}{(a_1 - a_4)(1 + a_2 e^{-a_3 t})} e^{-a_3 t}\right). \tag{4.20}$$

To follow the conventional model, $\alpha = 0$, the constants of Eq. 4.9 become:

$$a_1 a_2 = a_1 - a_4, \quad a_2 = 0, \quad \text{and} \quad a_3 = e_n. \tag{4.21}$$

Under these conditions, the effects of the shallow level are effectively ignored s.t. the fraction of electrons recaptured by the shallow level is zero. Thus, using Eq. 4.20, the transient photo capacitance becomes:

$$\Delta C(t) = \Delta C_{T0} \left(1 - e^{-e_n t}\right), \tag{4.22}$$

where $\Delta C_{T0} = C N_T / (2 N_D)$ is the saturation photo capacitance. Evidently, C_{T0} is equivalent to C_{ss} in the conventional model.

It can similarly be shown that for p-type material depicted in Fig. 4.2, with a net acceptor concentration, N_A, the trap concentration N is given by

$$N_o = \Delta N = 2 N_A \frac{\Delta C}{C}, \tag{4.23}$$

where C is the equilibrium reverse-bias capacitance, and ΔC is the capacitance change at the start of the saturating stimulation pulse.

4.2.2 Forward-Bias Photoresponses

Recently, the study of photoresponses has advanced sufficiently to permit measurements in the forward bias mode.

Figure 4.7 displays forward bias transient photo capacitance responses of a recent organic on p-silicon Schottky diode. The device is reportedly useful as both a photodiode and a photovoltaic cell [20]. The figure illustrates the significant impact of illumination on photo capacitance, indicating an initial increase in free carrier generation at the junction when illuminated. Upon turning off the illumination, the capacitance returns to its initial levels, indicating charge carrier trapping within deep impurity levels.

The difference between quasi-steady state and transient measurements in Schottky diodes can be attributed to the capacitive current (I_C), influencing the photo-generated current based on measurement conditions. Illumination charges diode capacitance, yielding a reduced diode current under a quasi-steady state, whereas turning off illumination discharges capacitance, resulting in a larger quasi-steady state current. This behavior can be described by [10]:

$$I_C = C \frac{dV}{dt} + V \frac{dC}{dt} \tag{4.24}$$

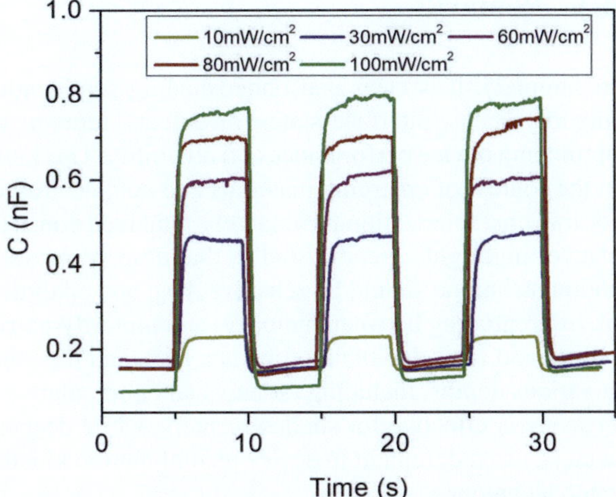

Fig. 4.7 Transient photo capacitance measured for an experimental p-silicon Schottky diode based on coumarin-doped graphene oxide [20]

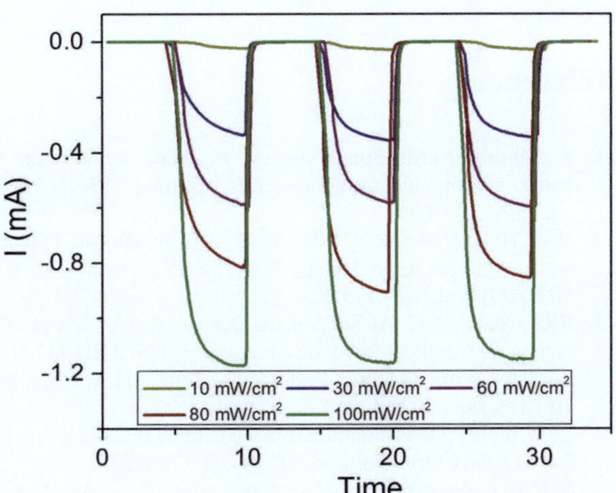

Fig. 4.8 Transient photocurrent for a sample p-silicon Schottky diode. Reproduced from Ocaya et al. [20]

Figure 4.8 shows the transient photocurrent responses of the same Schottky diode under pulsed illumination in forward bias. Under forward bias with quasi-steady state ($dV/dt = 0$), the equation simplifies to:

$$I_C(t) = -C_0 R_s \exp\left[b\frac{q}{kT}(V - IR_s)\right]\left(1 + b\frac{q}{kT}(V - IR_s)\right)\frac{dI}{dt} \qquad (4.25)$$

These example studies have generally aligned with recent modeling studies based on mathematical modeling [21, 22] and suggested that doping may degrade the performance of organic solar cells in many configurations.

4.3 Remarks

In summary, it is clear that understanding semiconductor traps, whether they are impurity states, interface states, or defects, remains of paramount importance for optimizing device performance and reliability. This knowledge is particularly crucial in the context of emerging materials and optoelectronic devices. While the methods for trap characterization have largely remained consistent over the past decade, this chapter highlights specific studies that deserve mention. The ideal trap characterization technique should be sensitive, fast, and analytically straightforward, capable of differentiating between majority- and minority-carrier traps. Spectroscopic capabilities and reproducible plotting are vital, and the method should encompass traps at various depths, including radiative and nonradiative centers. Luminescence spectroscopy is effective for shallow centers, while deeper nonradiative recombination centers, often dominant in device recombination kinetics, necessitate more sophisticated techniques.

References

1. S. Morab, M.M. Sundaram, A. Pivrikas, Review on charge carrier transport in inorganic and organic semiconductors. Coatings **13**(9), 1657 (2023). https://doi.org/10.3390/coatings13091657
2. H.-T. Huang, W. Cao, H.-H. Lin, Y.-C. Chin, GaAs$_{1-x}$Sb$_x$/GaAs single quantum well for long wavelength photonic devices. Solid State Electron. Lett. **1**(2), 98–104 (2019). https://doi.org/10.1016/j.ssel.2019.11.002
3. R.O. Ocaya, A.G. Al-Sehemi, A. Dere, A.A. Al-Ghamdi, F. Yakuphanoğlu, Electrical, photoconductive, and photovoltaic characteristics of a Bi2Se3 3D topological insulator based metal-insulator-semiconductor diode. Sens. Actuators A: Phys. **341**, 113575 (2022). https://doi.org/10.1016/j.sna.2022.113575
4. R. Williams, Determination of deep centers in conducting gallium arsenide. J. Appl. Phys. **37**, 3411 (1966). https://doi.org/10.1063/1.1708872
5. S. Heo, G. Seo, Y. Lee, M. Seol, S.H. Kim, D.-J. Yun, Y. Kim, K. Kim, J. Lee, Origins of high performance and degradation in the mixed perovskite solar cells. Adv. Mater. **31**(8), 1805438 (2019). https://doi.org/10.1002/adma.201805438
6. A.E. Khorasani, D.K. Schroder, T.L. Alford, A fast technique to screen carrier generation lifetime using DLTS on MOS capacitors, in *IEEE Transactions on Electron Devices*, vol. 61(9) (Institute of Electrical and Electronics Engineers (IEEE), 2014), pp. 3282–3288
7. M.K. Nazeeruddin, T.k. Ahn, J.K. Shin, Y.S. Kim, D.-J. Yun, K. Kim, J.-B. Park, J. Lee, M. Seol, Deep level trapped defect analysis in CH$_3$NH$_3$PbI$_3$ perovskite solar cells by deep level transient spectroscopy. Energy Environ. Sci. **10**(5), 1128–1133 (2017). https://doi.org/10.1039/C7EE00303J
8. C.T. Sah, L. Forbes, L.L. Rosier, A.F. Tasch Jr., Thermal and optical emission and capture rates and cross sections of electrons and holes at imperfection centers in semiconductors from photo and dark junction current and capacitance experiments. Solid-State Electron. **13**(6), 759–788 (1970). https://doi.org/10.1016/0038-1101(70)90064-X
9. D.V. Lang, Deep-level transient spectroscopy: a new method to characterize traps in semiconductors. J. Appl. Phys. AIP Publ. **45**(7), 3023–3032 (1974). https://doi.org/10.1063/1.1663719
10. G. Friesen, H.A. Ossenbrink, Capacitance effects in high-efficiency cells. Solar Energy Mater. Solar Cells **48**(1–4), 77–83 (1997). https://doi.org/10.1016/S0927-0248(97)00072-X

11. M. Soylu, M. Cavas, A.A. Al-Ghamdi, Z.H. Gafer, F. El-Tantawy, F. Yakuphanoğlu, Photo-electrical characterization of a new generation diode having GaFeO3 interlayer. Solar Energy Mater. Solar Cells **124**, 180–185 (2014). https://doi.org/10.1016/j.solmat.2014.01.045

12. E.H. Rhoderick, R.H. Williams, *Metal-semiconductor Contacts* (Clarendon Press, 1988)

13. E.H. Nicollian, A. Goetzberger, The Si–SiO, interface–electrical properties as determined by the metal-insulator-silicon conductance technique. Bell Syst. Tech. J. **46**(6), 1055–1033 (1967). Nokia Bell Labs

14. R.K. Gupta, F. Yakuphanoglu, Photoconductive Schottky diode based on Al/p-Si/SnS2/Ag for optical sensor applications. Solar Energy **86**(5), 1539–1545 (2012). https://doi.org/10.1016/j.solener.2012.02.015

15. İ Døkme, Ş Altındal, T. Tunç, İ Uslu, Temperature dependent electrical and dielectric properties of Au/polyvinyl alcohol (Ni, Zn-doped)/n-Si Schottky diodes. Microelectron. Reliab. **50**(1), 39–44 (2010). https://doi.org/10.1016/j.microrel.2009.09.005

16. O. Maida, D. Kanemoto, T. Hirose, Characterization of deep interface states in SiO_2/B-doped diamond using the transient photocapacitance method. Thin Solid Films **741**, 139026 (2022). https://doi.org/10.1016/j.tsf.2021.139026

17. S.K. Zhang, W.B. Wang, R.R. Alfano, H. Morkoç, Unusual transient photocapacitance in GaN/AlGaN multi-quantum wells. Superlattices Microstruct. **35**(1–2), 77–84 (2004). https://doi.org/10.1016/j.spmi.2004.03.070

18. P. Blood, J.W. Orton, The electrical characterization of semiconductors: majority carriers and electron states. Tech. Phys. (14), i–xxiii (1992)

19. A. Hierro, D. Kwon, S.A. Ringel, S. Rubini, E. Pelucchi, A. Franciosi, Photocapacitance study of bulk deep levels in ZnSe grown by molecular-beam epitaxy. J. Appl. Phys. **87**(2), 730–738 (2000). https://doi.org/10.1063/1.371933

20. R.O. Ocaya, A.G. Al-Sehemi, A. Al-Ghamdi, F. El-Tantawy, F. Yakuphanoğlu, Organic semi-conductor photosensors. J. Alloys Comp. **702**, 520–530 (2017). https://doi.org/10.1016/j.jallcom.2016.12.381

21. V.A. Trukhanov, V.V. Bruevich, D. Yu Paraschuk, Effect of doping on performance of organic solar cells. Phys. Rev. B **84**(20), 205318 (2011). https://doi.org/10.1103/PhysRevB.84.205318

22. G.F.A. Dibb, M.-A. Muth, T. Kirchartz, S. Engmann, H. Hoppe, G. Gobsch, M. Thelakkat, N. Blouin, S. Tierney, M. Carrasco-Orozco, others, Influence of doping on charge carrier collection in normal and inverted geometry polymer: fullerene solar cells. Sci. Rep. **3**(1), 3335 (2013). https://doi.org/10.1038/srep03335

Chapter 5
New Parameter Extraction Techniques

Abstract A simplified graphical method, termed the current-voltage-temperature (I-V-T) method, is proposed for extracting static parameters of Schottky diodes. This method extends the conventional current-voltage approach by incorporating temperature considerations. Initially, voltage-temperature characteristics are derived at two constant currents. Then, points at the same bias or temperature under specified current densities are used to extract the parameters of interest. This mathematical abstraction significantly streamlines the determination of ideality factor, barrier height, and diode resistance. The analysis starts by assuming an ideal diode with zero current modulation due to a base resistance, followed by quantifying the effect of a non-zero base resistance. The analysis includes simulations on the commercial BAT54 Schottky diode and on Au/n-Ge, Au/n-Si, and Au/n-GaAs Schottky diodes in the literature. The IVT method yields lower variances over wider bias in calculated barrier height, ideality factor, and forward resistance compared to the published values, suggesting that the I-V-T method is a feasible alternative for characterizing these diodes. This is primarily due to the ability to control current density in the TE equation.

5.1 Introduction

A Schottky diode is created when a junction forms between a metal and a semiconductor, creating a potential energy barrier referred to as a Schottky barrier [1–3]. The characterization of the Schottky barrier remains crucial, especially during device fabrication, where metallization of a semiconductor surface for external device connection may cause the formation of a potential barrier at the connection. However, characterizing it with respect to its static parameters can be challenging. A common parameter of interest is the Schottky barrier height (SBH), often deduced from transport measurements of barrier diodes created on non-degenerate semiconductors [4, 5]. Various measurement methods, including current-voltage (I-V), activation energy, capacitance-voltage (C-V), and photoelectric measurements, have been presented over the years [2]. These measurements are often made under simplifying assumptions that provide reasonable indications of expected parameter ranges and

R. Ocaya, *Extraction of Semiconductor Diode Parameters*,
https://doi.org/10.1007/978-3-031-48847-4_5

variations. For instance, the C-V method is carried out mostly in reverse bias using a high-frequency excitation signal, assuming that the diode won't exhibit a low-voltage resonance peak [4]. In the I-V method, it is essential to model the effect of the forward resistance of the diode [5, 6].

Existing models in the I-V extraction are broadly classified into three categories: simplistic models relying on thermionic emission theory, physical analytical models providing closed-form solutions, and numerical models based on Poisson's equation with drift-diffusion and continuity equations [5, 6]. Many engineering methods for processing I-V curves have restrictive assumptions about acceptable parametric values [5]. This chapter departs from these methods by examining the role of temperature more closely and showing that it can lead to better accuracy in the calculated static parameters [7, 8]. This approach requires a reinterpretation of existing I-V data, which already implicitly contains temperature behavior, into the constant current V-T format.

The primary aim of this chapter is to present an I-V-T method that is shown to simplify the graphical extraction of the static parameters, namely barrier height, ideality factor, and forward resistance of a Schottky diode. The method also leads to a reasonable estimation of the effective Richardson's constant of a diode of a specified cross-sectional area. We derive the mathematical basis starting at the diode current density and show how specific points on the constant current characteristics impact the calculation of the static parameters of interest. We begin by illustrating the method on results of an LT-Spice simulation of the commercial BAT54 Schottky diode [9, 10].

There are several valuable specifics with the BAT54, such as its operation at much higher currents and temperatures. Additionally, there is a manufacturer data sheet from which the static parameters can be estimated, with particular emphasis on the verifiable values of the dynamic resistance from its known room temperature I-V characteristic. The contemporary interest in metal-semiconductor junctions continues to generate much data in the literature for a wide range of Schottky barrier diodes, typically presented as I-V curves measured at different temperatures. For the analysis in this chapter, the numerical data were first extracted from published graphs into raw, equivalent tables and then converted to the constant current, voltage versus temperature format, which can be done using software digitizers such as Engauge [13].

The ease of data re-representation makes the method applicable to a vast number of arbitrary, new, and existing laboratory Schottky barrier diodes. As examples of potential application, we recalculate the static parameters for three experimental Schottky diodes constructed of gold and different semiconductors: Au/n-Si published by Sharma [12], Au/n-GaAs published by Singh et al. [11], and Au/n-Ge (100) published by Chawanda et al. [14]. Application of the I-V-T method to different regions of the characteristics produces much lower variances in calculated barrier height, ideality factor, and forward resistance when cited by their originators.

5.2 Theory

From thermionic emission theory [2, 5, 6], the ideal Schottky diode current density J can be written

$$J = A^* T^2 \exp\left[-q\Phi_b/(nkT)\right]\left(\exp\left[qV/(nkT)\right] - 1\right), \qquad (5.1)$$

where A^* is the effective Richardson's constant, Φ_b is the barrier height on the metal side, k is Boltzmann's constant, n is the ideality factor of the diode and T is absolute temperature. Suppose that the two current densities are in a known ratio a, such that

$$J_2 = aJ_1, \quad \text{for} \ \ 0 < a < 1. \qquad (5.2)$$

Figure 5.1 illustrates the *general* effect of the constant current densities J_1 and J_2 on the diode voltage as it is cooled [15, 16]. In the experiment, the diode is heated to a high initial temperature T_0 beyond point A in the figure under forward current density equal to J_1. The diode is then allowed to cool to T_1 where its forward voltage is V_{F1} (point B in Fig. 5.1). The current density is then switched to a new constant value J_2. A new terminal voltage V_{F2} is observed (point C). As the diode continues to cool down under the new current I_2 its terminal voltage will eventually reach V_F again, at which the measured temperature will be T_2. At temperature T_1 the current density becomes

$$J_1 \approx A^* T_1^2 \exp\left[-q\Phi_b/(nkT_1)\right] \exp\left[qV_F/(nkT_1)\right]. \qquad (5.3)$$

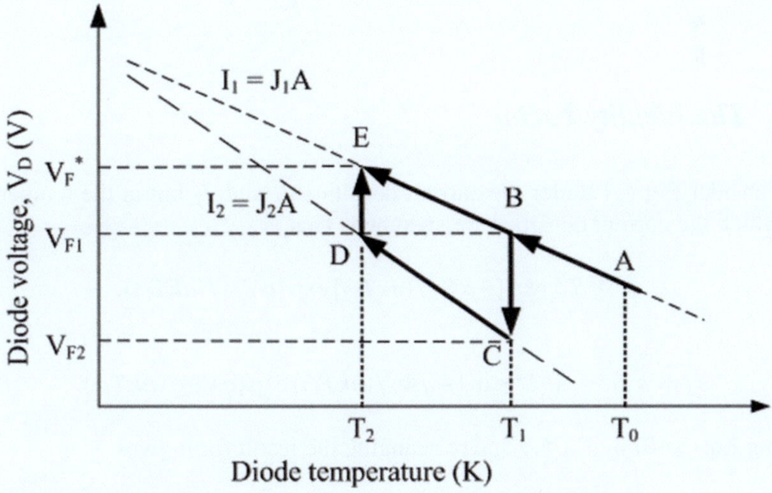

Fig. 5.1 Measurement trajectory (BCDE) applied to the Schottky diode

Similarly, at temperature T_2

$$J_2 = a J_1 \approx A^* T_2^2 \exp\left[-q\Phi_b/(nkT_2)\right] \exp\left[q V_F/(nkT_2)\right]. \tag{5.4}$$

5.2.1 Schottky Barrier Height

Dividing Eq. (5.4) by Eq. (5.3) gives

$$a = \left(\frac{T_2}{T_1}\right)^2 \exp\left[-\frac{q\Phi_b}{nk}\left(\frac{1}{T_2}-\frac{1}{T_1}\right)\right] \exp\left[\frac{q V_F}{nk}\left(\frac{1}{T_2}-\frac{1}{T_1}\right)\right]. \tag{5.5}$$

If we define a non-zero temperature T^* such that

$$\frac{1}{T^*} = \frac{1}{T_2} - \frac{1}{T_1}, \tag{5.6}$$

then Eq. (5.5) can be written, after rearranging, as

$$a\left(\frac{T_1}{T_2}\right)^2 = \exp\left[\frac{q(V_F - \Phi_b)}{nkT^*}\right]. \tag{5.7}$$

Hence the barrier height can be found from

$$\Phi_b = V_F - \frac{2nkT^*}{q} \ln\left(\frac{T_1}{T_2}\sqrt{a}\right). \tag{5.8}$$

5.2.2 The Ideality Factor

Now consider Fig. 5.1 under the current densities J_1 and J_2 but at the temperature T_1 at which the current densities are switched. That is

$$J_1 \approx A^* T_1^2 \exp\left[-q\Phi_b/(nkT_1)\right] \exp\left[q V_{F1}/(nkT_1)\right], \tag{5.9}$$

and

$$J_2 = a J_1 \approx A^* T_1^2 \exp\left[-q\Phi_b/(nkT_1)\right] \exp\left[q V_{F2}/(nkT_1)\right]. \tag{5.10}$$

Dividing Eq. (5.10) by Eq. 5.9 and rearranging the result for n gives

$$n = \frac{q\Delta V(T_1)}{kT_1 \ln a}, \tag{5.11}$$

where $\Delta V(T_1)$ is the observable voltage difference $(V_{F2}\text{-}V_{F1})$. This is precisely the result that was obtained using similar methods for the p-n diode [15, 16]. Substituting Eq. (5.11) into Eq. (2.1) gives the barrier height in terms of current ratio, voltage and temperature as

$$\Phi_b = V_F - 2\frac{T_2}{T_1 - T_2}\left[\frac{\ln(T_1\sqrt{a}/T_2)}{\ln a}\right]\Delta V(T_1). \tag{5.12}$$

5.2.3 Series Resistance

In the presence of series diode resistance R, the current density is modified by the term IR. If we rewrite Eq. (5.1) as the actual current flowing through the cross-sectional area A of the diode, then

$$I \approx A^*AT^2 \exp\left[-q\Phi_b/(nkT)\right]\exp\left[q(V - IR)/(nkT)\right]. \tag{5.13}$$

It follows that at temperature T_1, under current density J_2 one can write

$$J_2 \approx A^*T_1^2 \exp\left[-q\Phi_b/(nkT_1)\right]\exp\left[q(V_{F2} - I_2R)/(nkT_1)\right]. \tag{5.14}$$

Similarly, at temperature T_2, under current density J_2

$$J_2 \approx A^*T_2^2 \exp\left[-q\Phi_b/(nkT_2)\right]\exp\left[q(V_{F2} - I_2R)/(nkT_2)\right], \tag{5.15}$$

where $I_2 = J_2 A$. Dividing Eq. (5.14) by Eq. (5.15) and simplifying gives

$$\left(\frac{T_1}{T_2}\right)^2 \approx \exp\left(-\frac{q\Phi_b}{nkT^*}\right)\exp\left(\frac{V_{F2}}{nkT_1}\right)\exp\left(-\frac{V_{F1}}{nkT_2}\right)\exp\left(\frac{I_2R}{nkT^*}\right). \tag{5.16}$$

Similarly, at the same temperatures as before but with a current density J_1 one can write

$$\left(\frac{T_1}{T_2}\right)^2 \approx \exp\left(-\frac{q\Phi_b}{nkT^*}\right)\exp\left(\frac{V_{F1}}{nkT_1}\right)\exp\left(-\frac{V_F^*}{nkT_2}\right)\exp\left(\frac{I_1R}{nkT^*}\right), \tag{5.17}$$

where $I_1 = J_1 A$. It is evident that $I_2 = aI_1$. Since Eqs. (5.16) and (5.17) express the same result, they can be equated. Comparing and simplifying the exponents of the resulting equation give

$$-\frac{V_{F1}}{T_1} + \frac{V_F^*}{T_2} + \frac{V_{F2}}{T_1} - \frac{V_{F1}}{T_2} = \frac{R}{T^*}(I_1 - I_2). \tag{5.18}$$

Hence, the series resistance can be found:

$$R = \left(\frac{T^*}{I_1 - I_2}\right)\left[\frac{V_F^* - V_{F1}}{T_2} - \frac{V_{F2} - V_{F1}}{T_1}\right]. \tag{5.19}$$

A simple argument against ignoring the series resistance is as follows. Considering any two I-V-T characteristics at a given temperature, it follows that when the current density is switched from J_1 to J_2 the observed diode voltage falls from V_1 to V_2. Hence, using Ohm's law, the dynamic resistance at a given temperature T_s is

$$R = \left.\frac{dV}{dI}\right|_{T_s} \approx \left.\frac{\Delta V}{\Delta I}\right|_{T_s} \approx \left.\frac{V_1 - V_2}{I_1 - I_2}\right|_{T_s}. \tag{5.20}$$

For example, typical data based on [12] (see Table 5.3) gives

$$R \approx \left.\frac{(0.798 - 0.665)\,V}{(10 - 1)\mu A}\right|_{310K} \approx 15.3k\Omega. \tag{5.21}$$

This is the typical value calculated for the referenced diodes below. The diode voltage reduction by the term IR in Eq. (5.13) at $10\mu A$ is then 0.153 V, which is appreciable when compared to the diode voltage at that temperature. However, the *general* trend is a voltage differential that drops gradually with temperature at a constant current differential. This signifies that the calculated resistance will also drop with temperature, a trend that has also been observed in the calculated resistances of the referenced diodes.

It is also evident that the referenced diodes are generally low-current devices by virtue of their high dynamic resistance, although the exact resistance variation with current is not in the present scope of this chapter. Nonetheless, as in the case of the p-n diode, the dynamic resistance is expected to be a function of the current at a given temperature. This is based on the observation in Figs. 5.8, 5.9 and 5.10 that if J_2 is increased then the voltage difference does not increase as fast as the current. Hence the resistance is somewhat inversely proportional to diode current. Table 5.5 shows the calculated barrier heights with standard deviations, against literature values for gold-based Schottky diodes in the literature.

5.2.3.1   Regions of Validity

The requisite condition for the validity of the foregoing method is linearity in the I-V-T characteristics in the neighborhood of the adjacent temperature points, T_1 and T_2. A preemptive inspection of the result Figs. 5.8, 5.9 and 5.10 suggests that this is roughly the case for the referenced diodes over the investigated temperature range. This condition may be investigated by finding the region where the differential diode voltage, ΔV, is a linear function of temperature. This differential is taken as the difference in diode voltage at the same temperature but under the two currents. For the referenced experimental Schottky diodes the two currents used are $1\,\mu A$ and $10\,\mu A$, and the temperature range is restricted to about 140 to 300 K. Section 5.5 presents the mathematical basis of the method and justifies the requisite condition of linearity.

5.2.4 Measurement Quality and Uncertainty

The measurement quality of a given measured result depends on both the actual error in the measurement process and the known structure of the error. The present method reduces the determination of the barrier height to two essential measurements, namely voltage, and temperature, which can be done with high precision and repeatability. Therefore the quality of the uncertainty estimate of the barrier height can be improved by the enhanced knowledge of the error structure. However, increased knowledge of the error structure can lead to a higher, albeit more realistic estimated uncertainty and consequently higher measurement quality. Practical measurements using this method should ideally not rely on the transcribing of voltage-temperature data points from a data sheet. Therefore, the method indicates the need for a new instrument that effectively maintains the current through a diode constant as its temperature is varied and the diode voltage is measured over the experimental range. Such an instrument remains in the clear scope of our future work. Practical measurements can have many possible sources of uncertainty [18].

In this chapter, the main measurement is the SBH. It is well-defined within the TE model. The method of acquiring samples for analysis is well-defined and is typical of characterization measurements for diodes i.e. I and V are measured at a point where T is known. The environmental conditions and their uncertainties are therefore well expressed for each measurement. We expect that the data itself does not exhibit personal bias because of the method used to acquire it. For instance, in the case of the simulations, the data were recorded directly as displayed. Similarly, the graphically published experimental data were obtained using special I-V characterization equipment under controlled conditions of temperature. Therefore, the most likely sources of uncertainty in the proposed method are the presumed linearity of the model and the instrument measurement error for temperature and voltage measurement.

A direct investigation of the error distributions in an actual measurement setting can be done starting from Eq. (5.12) under the necessary condition that the two diode currents are constant. In practice, precision constant current generators exist or can readily be constructed to meet this condition. Equation (5.12) may then be written as

$$\Phi_b = V_F - \alpha \frac{T_2}{T_1 - T_2}(V_1 - V_2) \ln \frac{\beta T_1}{T_2}, \tag{5.22}$$

where $\alpha = 2/\ln a$ and $\beta = \sqrt{a}$ are constant. The error structure can be deduced from Eq. (5.36) for $\Phi_b = \Phi_b(V_F, V_1, V_2, T_1, T_2)$ i.e.

$$\partial \Phi_b \approx \sqrt{\sum_{i=1}^{5}\left(\left.\frac{\partial \Phi_b}{\partial x_i}\right|_{x_i,best}\delta x_i\right)^2}. \tag{5.23}$$

Differentiating Eq. (5.22) with respect to each of the five variables in the measurement set gives:

$$\frac{\partial \Phi_b}{\partial V_F} = 1$$

$$\frac{\partial \Phi_b}{\partial V_1} = -\frac{\partial \Phi_b}{\partial V_2} = -\frac{\alpha T_2}{T_1 - T_2}\ln\frac{\beta T_1}{T_2}$$

$$\frac{\partial \Phi_b}{\partial T_1} = -\alpha\frac{V_1 - V_2}{T_1 - T_2}\left(\frac{T_2}{T_1} - \frac{T_2}{T_1 - T_2}\ln\frac{\beta T_1}{T_2}\right) \tag{5.24}$$

$$\frac{\partial \Phi_b}{\partial T_2} = \alpha\frac{V_1 - V_2}{T_1 - T_2}\left(1 - \frac{T_1}{T_1 - T_2}\ln\frac{\beta T_1}{T_2}\right).$$

Using Eqs. (5.24) with the known measurement uncertainties in the five variables allows the direct calculation of the worst-case $\Delta \Phi_b$ for a given measurement of Φ_b. Typical diode voltage and temperature measurements using the Keithley-230 electrometer [22] and a type J-thermocouple (iron-constantan) based digital thermometer [30] can have uncertainties of $\pm 0.05\%$ of the voltage reading and 0.5K for the temperature respectively. Applying these uncertainties and Eqs. (5.24) into Eq. 5.23) for the typical results shown in Table 5.1 gives an uncertainty of $\Delta \Phi_b = \pm 0.007$V.

If one assumes a reasonably higher uncertainty, such as can be expected in a deduction of data points from a graph i.e. at 10% of the reading for both voltage and temperature measurements, then recalculation using the foregoing equations gives $\Delta \Phi_b = \pm 0.2$V. This value is within the range of the calculated uncertainties of the final results, Table 5.5 below, that were deduced graphically. In practice, the measurement accuracies of voltage and temperature are expected to be much better than 10%, as indicated by the Keithley instrument and the digital thermometer. The potential contribution of non-linearity to the uncertainty in SBH is investigated in Sect. 5.5. Equation 5.6 describes how the uncertainties stated in the result tables below were calculated.

5.2.5 Simulation Approach

The LTSpice simulation circuit comprising a single current source in series with the diode is shown in Fig. 5.2. Practical implementations of the current source are covered extensively in [15, 16]. The LTSpice command that achieves the simulation is

```
.dc temp -60 130 1 I1 1m 10m 9m
```

Fig. 5.2 LTSpice [9]
simulation circuit used to
investigate the I-V-T
behavior of the BAT54
Schottky barrier diode

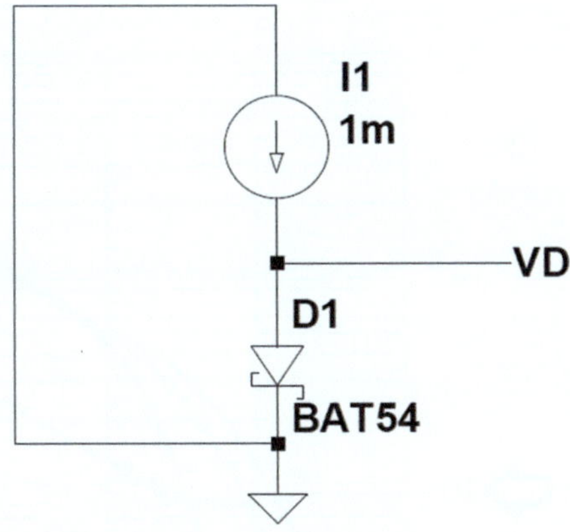

5.2.6 *Experimental*

The three Schottky diodes referenced and analyzed were constructed differently. In
[12], the n-region comprised an $8.2\,\mu$m thick epitaxial n-layer with of 1.35×10^{15}cm^{-3}
concentration. Ohmic contact was made by depositing a 200 nm thick aluminium (Al)
layer. A gold (Au) film of thickness 80 nm was deposited on the epitaxial n-layer. In
[11] the n-GaAs epitaxial layer was deposited over Si-GaAs by metal-organic chemical vapor deposition (CVD). The silicon (Si) dopant concentration in the epitaxial
layer was 1.0×10^{16} cm^{-3}. An ohmic Au contact of 2mm diameter and 100 nm thickness was deposited. The Schottky contact also consisted of Au of 2 mm diameter and
100 nm thickness on the same side as the ohmic contact. In [14], the Au/n-Ge (100)
Schottky barrier diodes were fabricated on antimony (Sb)-doped (100) n-type Ge
substrate with a doping concentration of 2.5×10^{15}/cm^3. Inside a vacuum chamber,
a 100 nm thick AuSb back ohmic contact was deposited by resistive evaporation.
Schottky contacts of Au were also deposited by vacuum resistive evaporation. The
contacts measured at 0.60 ± 0.05 mm diameter and 30 nm thickness. Current-voltage
measurements were conducted using Keithley model-230 and model-617 electrom-
eters at ± 0.5K temperature resolution.

5.3 Typical Results and Analysis

The basic idea is to re-represent I-V characteristics in the I-V-T form. Simply put all
the points on the I-V characteristics at the same current have their voltages collected
at the temperature of the characteristic curve as in Fig. 5.3. In this way, the specified
current is essentially constant with a peculiar V-T characteristic.

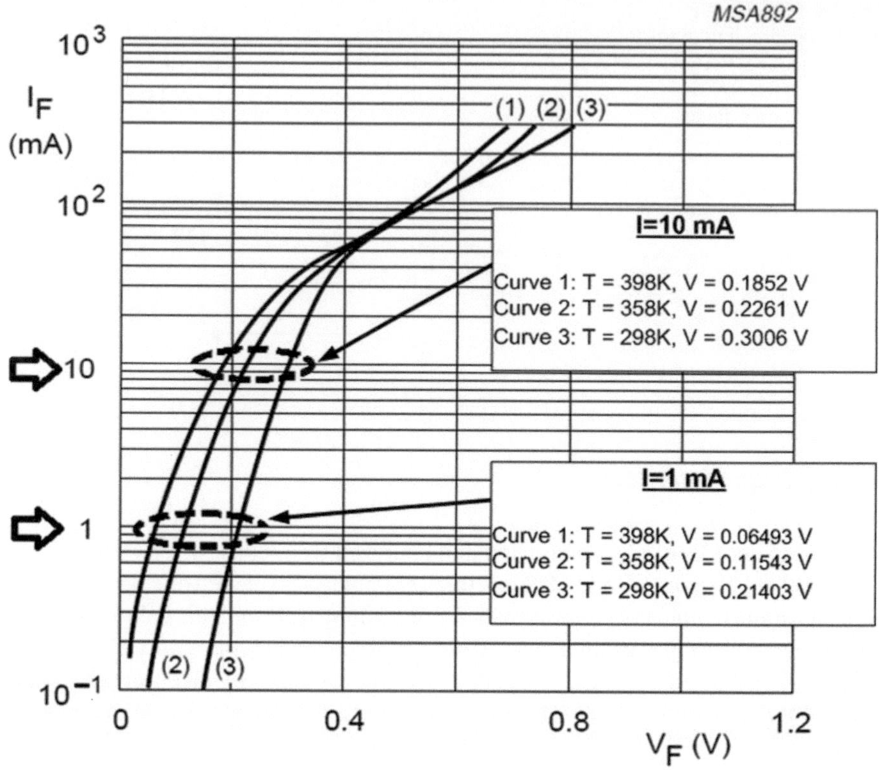

Fig. 5.3 BAT54 Schottky diode data sheet [10] I-V responses showing iso-current points at 1 and 10 mA (encircled) for I-V-T form determination

5.3.1   BAT54 Diode Simulation Results

Figure 5.4 plots the LTSpice simulated I-V-T response together with the I-V-T data extracted manually from the BAT54 datasheet [10] over the same temperature range. The worst case error in the diode voltage is about 10% at the lower temperature end-points but clearly appears to increase as the temperature falls. This disagreement is likely to be a limitation of the LTSpice model of the device but is within acceptable bounds for the illustrative purposes of this chapter. Table 5.1 shows the output of the constant-current V-T simulation. To build up the data in Table 5.1 a cursor on the LTSpice graphical user interface was placed at the temperature points corresponding to Fig. 5.1. The voltage and temperature at that point were then recorded off the displayed coordinates of the cursor. The number of significant figures in Table 5.1 for both voltage and temperature were the default for the simulation.

In practice, the measurement of diode voltages will be generally repeatable to 3 significant digits when measured with a digital multimeter [15]. The use of a 10-bit

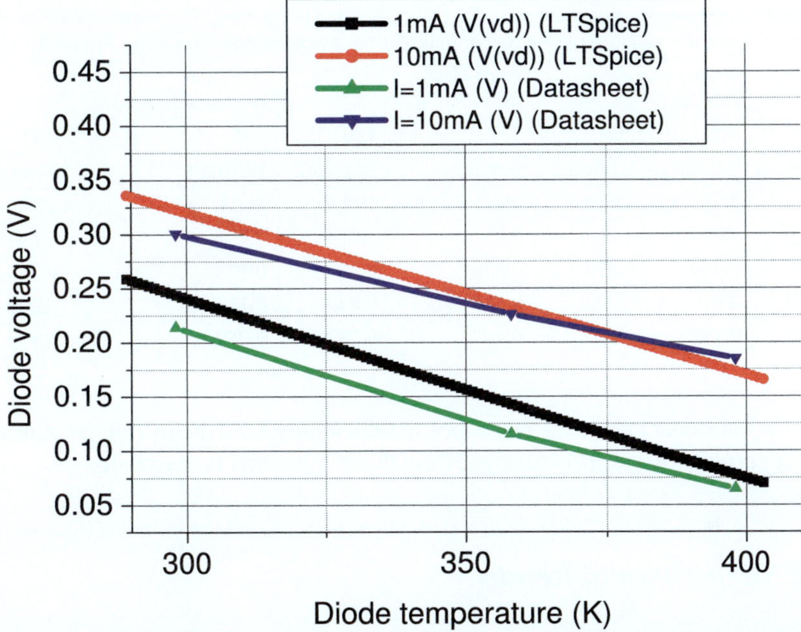

Fig. 5.4 A graph showing the converted I-V-T form of the BAT54 Schottky diode datasheet [10] at 298, 358, and 398 K versus a comparable LTSpice simulation

Table 5.1 Graphically determined V-T values from the LTSpice simulation around 298 K. The significance of each parameter is shown in Fig. 5.1

Parameter	Value
T_1	298.18 K
T_2	250.38 K
V_F^*	0.3928 V
V_{F1}	0.3231 V
V_{F2}	0.2435 V

Table 5.2 Summary of calculated results using the method on the LTSpice data of Table 5.1

Parameter	Value
Φ_b	0.677 V
n	1.347
R	1.94 Ω

analog-to-digital converter (ADC) for computer-based data acquisition may have greater repeatability and convenience although a specification of the measurement uncertainties based on the bit-resolution will not necessarily be better [23].

Table 5.2 reports the resistance of the BAT54 using data from LTSpice in the method as approximately 2Ω. An estimate of the dynamic resistance of the BAT54

Table 5.3 Summary of values determined graphically using the method. The parameter T_C indicates the center of the temperature measurement neighborhood obtained by averaging T_1 and T_2

T_C/K	Au/n-Si (Sharma)		Au/n-GaAs (Singh et. al.)		Au/n-Ge (Chawanda et. al.)	
	280	208	277	180	245	205
Parameter	Region 1	Region 2	Region 1	Region 2	Region 1	Region 2
T_1/K	310.1	250.0	300.1	199.9	260.0	220.0
T_2/K	250.4	165.9	245.2	160.4	230.5	190.7
V_F^*/V	0.474	0.898	0.656	0.924	0.178	0.274
V_{F1}/V	0.316	0.798	0.527	0.849	0.114	0.202
V_{F2}/V	0.179	0.665	0.407	0.732	0.057	0.138

using the datasheet [10] gives resistance in the vicinity for a diode voltage of around 0.3V at currents of 1 to 10mA, suggesting that the method is reasonable.

5.3.2   *Experimental Results*

With respect to the experimental referenced diodes [11, 12, 14], only the data in the approximate range of 140–350 K at the two currents of 1μA and 10μA were considered for reasons of fair comparison. Figures 5.5, 5.6 and 5.7 reproduce the I-V characteristics of those experiments. Figures 5.8, 5.9 and 5.10 show the re-representation

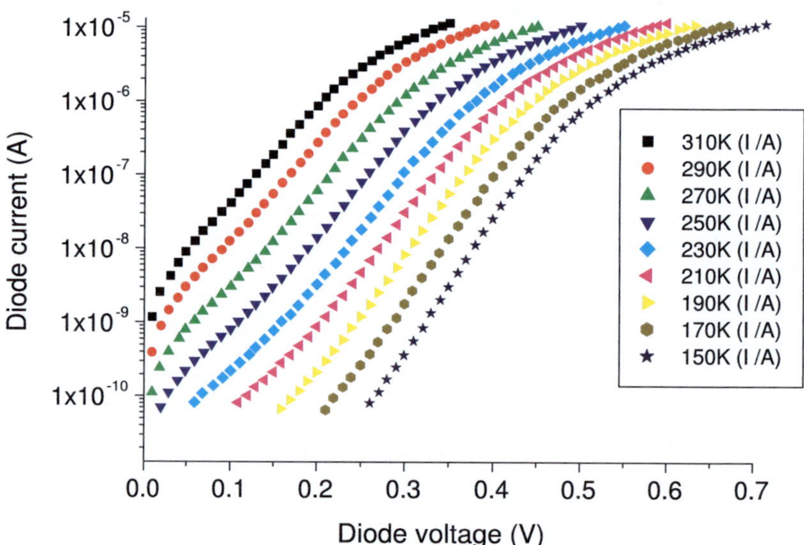

Fig. 5.5 Au/n-Si experimental barrier diode data [12]

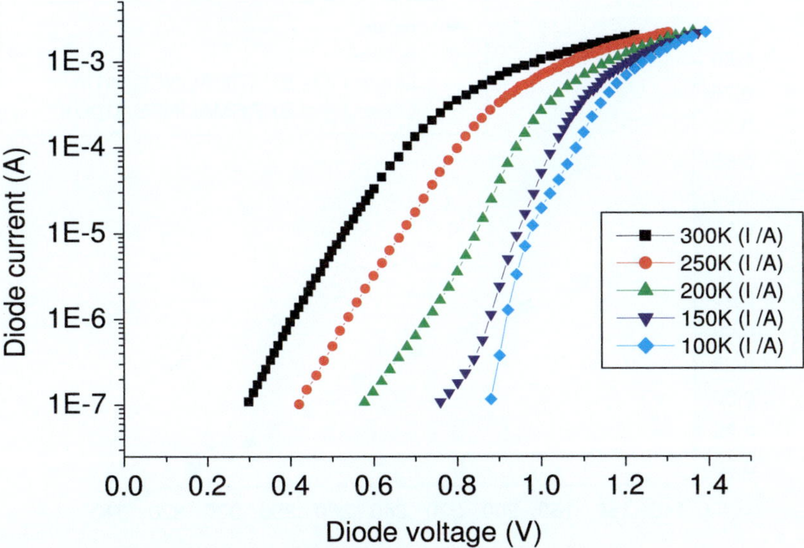

Fig. 5.6 Au/n-GaAs experimental barrier diode data [11]

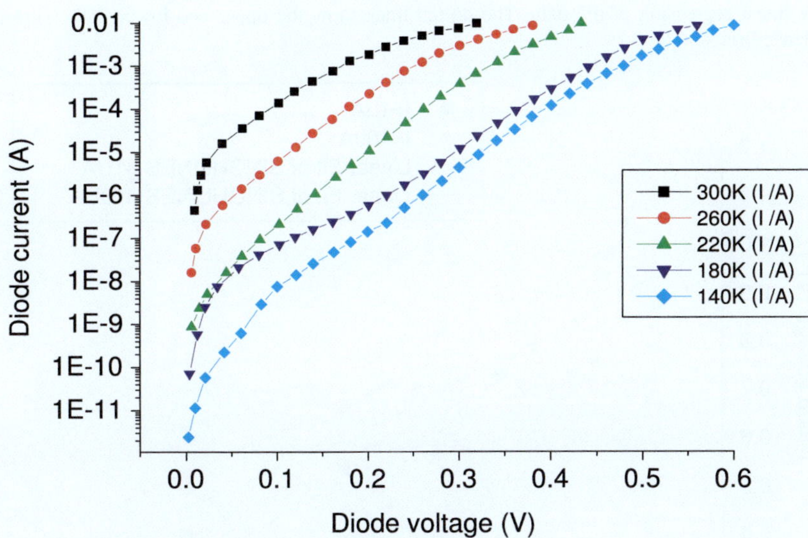

Fig. 5.7 Au/n-Ge experimental barrier diode data [14]

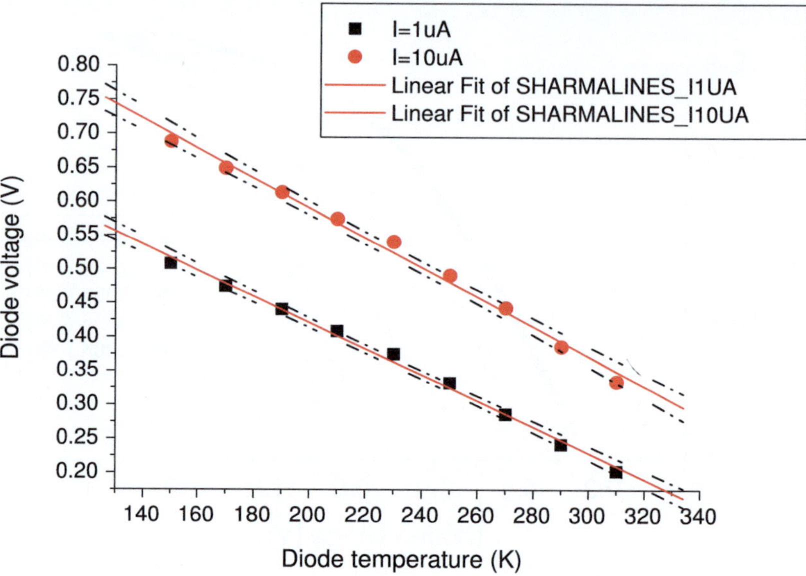

Fig. 5.8 Au/n-Si Schottky barrier diode experimental results based on [12]. The solid lines (red) are the linear regression of the data. The dotted lines show the upper and lower 95% regression confidence limits

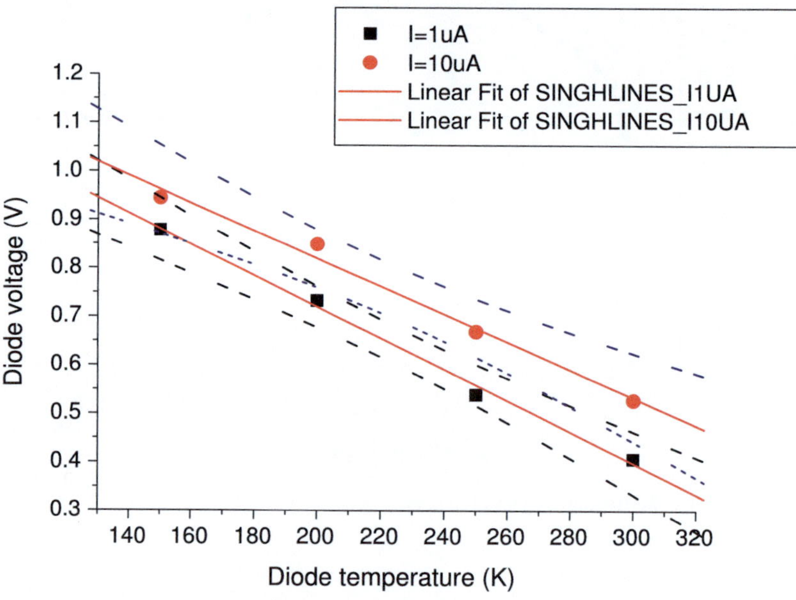

Fig. 5.9 Au/n-GaAs Schottky barrier diode experimental results based on [11]. The solid lines (red) are the linear regression of the data. The dotted lines show the upper and lower 95% regression confidence limits

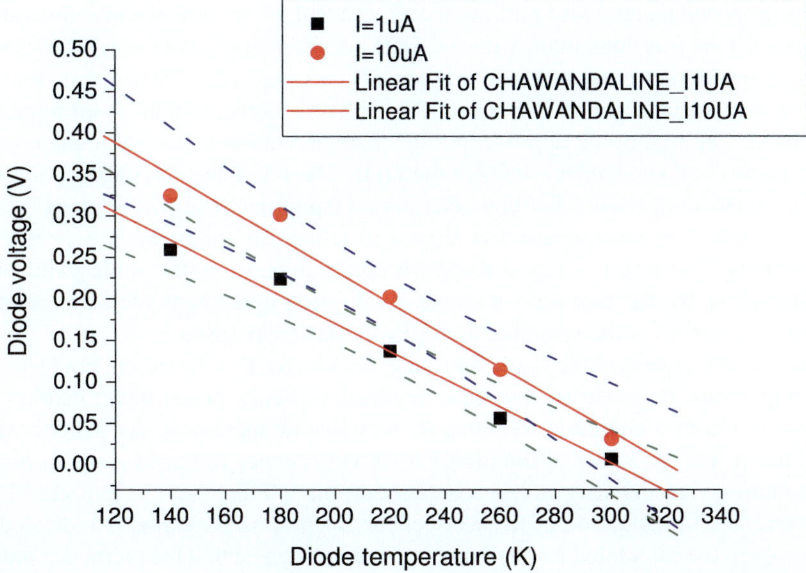

Fig. 5.10 Au/n-Ge (100) Schottky barrier diode experimental results based on [14]. The solid lines (red) are the linear regression of the data. The dotted lines show the upper and lower 95% regression confidence limits

of the experimental I-V characteristics as V-T curves at the constant currents of 1 and $10\,\mu$A. The solid lines (red) show the linear regression of the constant-current V-T data. The dotted lines show the upper and lower 95% confidence limits of the linear regression.

5.4 IVT Measurements

We have proposed an I-V-T method to determine the barrier height, ideality factor, and forward resistance static parameters of an arbitrarily specified diode for which I-V characteristics exist. From these parameters, subsequent analysis can reveal other important parameters such as the Richardson constant, although this was not the primary concern of this chapter. The proposed I-V-T method begins by exploring the effect of maintaining constant diode current density on the voltage and temperature response. It then proves mathematically that the calculation of the barrier height, ideality factor, and resistance are simplified. We show how this is possible using the proposed I-V-T method since it reduces an unwieldy and generally difficult-to-handle current density expression to a more manageable one. In essence, the effect of temperature on the static parameters of interest is examined more closely and is shown to assist calculations.

The proposed method also outlines a way to calculate the variance in a given static parameter. This was done mainly for the SBH over the temperature range 140-350K, the range over which comparative literature data was available. The proposed method exhibits low variance in the SBH and Richardson's constant that is subsequently calculated. The traditional approach to the characterization of the static parameters has been through capacitance-voltage methods. The I-V-T has the strong point that existing multi-temperature I-V characterization equipment can still be used, except that the method of interpretation is altered to reflect the temperature aspect more significantly. The results suggest that with minimal exertion, the static parameters as determined by the proposed method are in good agreement with independent published sources for the specified Schottky diode construction.

There is still considerable scope for future work on the I-V-T method. For instance, the temperature response of the static parameters needs better quantification for the low and high-temperature regions. At very low temperatures, for instance, it is well known that the effect of the image force is to reduce the SBH [24]. At higher temperatures, non-linearity in the constant current V-T behavior is expected [20]. Therefore, the investigation of these effects on the proposed method is indicated.

However, the calculated low variances appear to suggest that the method is indeed suitable in a very general sense over the wide range examined. Also, as highlighted by simulation and data sheet interpretation of the commercial, higher current, and higher temperature BAT54 device, the method is general enough to be used with an arbitrarily defined Schottky barrier diode. Finally, the potential for automating the I-V-T method opens up new possibilities for suitably designed low-cost research instrumentation.

5.5   Derivations

The foregoing method presumes linearity of the constant current V-T characteristics in the neighborhood of the adjacent temperature points, T_1 and T_2. This presumption can be verified using either one of two approaches that are based on the observation that both $\partial I/\partial T$ and $\partial I/\partial V$ vanish under constant current density. We begin by rewriting Eq. 5.13) as

$$I \approx A'T^2 e^{\beta(V-\Phi_b-IR)/T}, \tag{5.25}$$

where $A'=A^*A$ and $\beta=q/(nk)$. Differentiating Eq. 5.25 first w.r.t temperature and then diode voltage gives

$$\frac{\partial I}{\partial T} \approx A'e^{\frac{\beta(V-\Phi_b-IR)}{T}}\left\{\beta T\left(\frac{\partial V}{\partial T} - \frac{\partial \Phi_b}{\partial T} - R\frac{\partial I}{\partial T}\right) - \beta(V - \Phi_b - IR) + 2T\right\}, \tag{5.26}$$

and

$$\frac{\partial I}{\partial V} \approx A'e^{\frac{\beta(V-\Phi_b-IR)}{T}} \left\{ \alpha\left[T\left(1 - \frac{\partial \Phi_b}{\partial V} - I\frac{\partial R}{\partial V} - R\frac{\partial I}{\partial T}\right)\right.\right.$$
$$\left.\left. - (V - \Phi_b - IR)\frac{\partial T}{\partial V}\right] + 2T\frac{\partial T}{\partial V}\right\}. \quad (5.27)$$

In the first approach, take $\partial I/\partial T = 0$ and assume temperature independence of barrier potential. Equation (5.26) then becomes

$$\frac{\partial V}{\partial T} - \left(\frac{V - \Phi_b - IR}{T}\right) + \frac{2}{\beta} \approx 0. \quad (5.28)$$

Hence

$$V \approx \left(\frac{\partial V}{\partial T} + \frac{2}{\beta}\right)T + \Phi_b + IR. \quad (5.29)$$

Equation 5.29 may appear unusual but it serves to emphasize that the gradient of the V-T plot is modified by the ideality factor (contained in β). This behavior is observable in many diode experiments [15, 24]. For the diodes typically studied,

$$\alpha = \left|\frac{\partial V}{\partial T}\right| >> \frac{2nk}{q} \quad (5.30)$$

so that

$$V \approx \alpha T + \Phi_b + IR, \quad \text{where } \alpha < 0. \quad (5.31)$$

Equation 5.31 permits the estimation of Φ_b and its uncertainty $\Delta\Phi_b$ using the independent experiment variables V and I and their uncertainties at the arbitrary temperature T.

The second approach can be illustrated as follows. The SBH is known to increase with forward bias voltage V [25–27], i.e.

$$\Phi_b(V) = \Phi_{bo} + \gamma V, \quad \text{with } \frac{\partial \Phi_b}{\partial V} = \gamma > 0, \quad \text{and } n = \frac{1}{1 - \gamma}, \quad (5.32)$$

where Φ_{bo} is the zero-bias barrier height. Under constant current and constant diode resistance [32], we may write $\partial I/\partial V = \partial R/\partial V = \partial I/\partial T = \partial R/\partial T = 0$. Equation 5.27 then reduces to

$$T\left(1 - \frac{\partial \Phi_b}{\partial V}\right) - (V - \Phi_b - IR)\frac{\partial T}{\partial V} + \frac{2T}{\beta}\frac{\partial T}{\partial V} = 0. \quad (5.33)$$

Using the relations in Eq. 5.32 into Eq. (5.33) and simplifying reduces to Eq. (5.31).

5.5.1   Deviations from the Ideal TE Model

Deviations from the ideal TE model at arbitrary experimental diode voltages, temperatures and constant currents can be estimated as follows. First, a point on the V-T characteristic at which it is most linear is found along with the best gradient at that point. Hence Φ_{bo} can then be found. By using Φ_{bo}, the gradient at any other V and T can be calculated from Eq. 5.31. A comparison of the different gradients as a function of temperature will give an indication of the deviation from the "perfect" straight line of the ideal TE model. The effect of non-linearity as a result of temperature [20] is an interesting direction for future work on the method.

5.6   Evaluation of Standard Uncertainty

5.6.1   Estimating Instantaneous Uncertainty in SBH

Eq. 5.31 implies that for a given diode resistance and measurement temperature, a smaller current is generally beneficial to the measurement. The term "instantaneous" above means the specific measurement points of I, V, and T. Figure 5.1 shows that the *magnitude* of the slope of the V-T characteristic is higher for smaller currents. For the experimental data surveyed, a tenfold increase in current produced a change with a factor of about 1.1 on both slope and ideality factor using Eq. 5.11. The ratio $(\partial V/\partial T)/n$ in Eq. 5.31 is therefore not significantly affected by changes in forward current. Following these arguments, Eq. 5.31 may be written in the form:

$$\Phi_b \approx V + \alpha T + IR \tag{5.34}$$

with a change in sign for α for maximum error propagation. For a linearly regressed model of a given function f of n variables, i.e.

$$f = \sum_{i=1}^{n} \rho_i x_i \tag{5.35}$$

with certain coefficients ρ, the rule of propagation of multiple independent uncertainties [17] is

$$\partial f \approx \sqrt{\sum_{i=1}^{n} \left(\left. \frac{\partial f}{\partial x_i} \right|_{x_i,best} \delta x_i \right)^2}. \tag{5.36}$$

Applying Eqs. 5.36 to (5.34) leads to

$$\partial\Phi_b \approx \sqrt{(\delta V)^2 + (\alpha\,\delta T)^2 + (R\delta I)^2}. \tag{5.37}$$

It is the control of the current that gives this method its linear character i.e. $\delta I = 0$. Consequently, at any given measurement point the propagated uncertainty in SBH can be written as the experimental uncertainty:

$$\Delta\Phi_b \approx \sqrt{(\Delta V)^2 + (\alpha\,\Delta T)^2}. \tag{5.38}$$

5.6.2 Standard Deviation and Confidence Limits

Standard uncertainties can be evaluated in one of two ways depending on whether or not the measurand is the result of several repeated measurements, giving rise to type "A" and type "B" evaluations respectively [18, 19]. This categorization is not meant to suggest that one is better than the other in a given analysis, but that it may rather be a matter of convenience or preference in a given analysis. In any case, both types rely on probability distributions, and the uncertainties are still governed by variances and standard deviations. The calculated errors in Table 5.5 are expressed as a standard deviation of the parameter. They were found by first applying Eq. 5.38 to the temperature regions shown in Table 5.4 for each device type. The standard deviation was then calculated. Finally, the 95% confidence intervals of all the linear regressions were calculated under the assumption of a normalized distribution using MicroCal Origin software and plotted, as shown in Figs. 5.8, 5.9 and 5.10.

Table 5.4 Summary of results calculated using the method on the data of Table 5.3 in the different temperature regions

	Au/n-Si [12]		Au/n-GaAs [11]		Au/n-Ge [21]	
T_C /K	280	208	277	180	245	205
Parameter	Region 1	Region 2	Region 1	Region 2	Region 1	Region 2
Φ_b/V	0.781	0.967	1.098	1.230	0.514	0.572
n	2.233	2.682	2.023	2.930	1.109	1.480
R /kΩ	27.2	3.9	19.5	9.6	12.7	13.3

Table 5.5 Comparison of the results calculated using the proposed I-V-T method against various literature sources with comparable donor concentration ($\approx 10^{-15}$/cm^3). The abbreviation 'nc' = not calculated

Literature source	$\Phi_b \pm \Delta\Phi_b$ (eV)	$n \pm \Delta n$	A** ($\times 10^6$ A/m^{2}K^2)	Method
Au/n-Si Schottky diode				
1. Sharma [12]	1.02±nc	$\approx$1.25±nc	0.9±nc	C-V
2. Sze,-Ng,-Cowley [2, 3]	0.82±0.10	–	1.12±nc	C-V
3. Proposed	0.874±0.130	2.457±0.320	1.02±0.30	I-V-T
Au/n-GaAs Schottky diode				
1. Singh et al. [11]	0.82±nc	1.9±nc	0.1±nc	C-V
2. Sze,-Ng,-Cowley [2, 3]	0.87±0.05	–	–	C-V
3. Hudait & Krupanidhi [31]	0.893±nc	2.95±nc	0.08±nc	I-V-T
4. Proposed	1.164±0.210	2.312±0.41	0.116±0.059	I-V-T
Au/n-Ge Schottky diode				
1. Chawanda et al. [21, 28]	0.615±0.086	$\approx$1.10±nc	0.137±nc	C-V
2. Chawanda et al. [14, 28]	0.529±0.019	1.413±0.270	–	C-V,I-V
3. Yao et al. [29]	0.503±0.006	1.4±nc	0.5±nc	C-V
4. Proposed	0.543±0.040	1.295±0.260	0.410±0.03	I-V-T

References

1. H.M. Manohara, E.W. Wong, E.Schlecht, B.D. Hunt, P.H. Siegel, Carbon nanotube schottky diodes using Ti-Schottky and Pt-Ohmic contacts for high frequency applications. Nano Lett. **5**(7), 1469–1474 (2005)
2. S.M. Sze, K.K. Ng, *Physics of Semiconductor Devices*, 3rd ed. (Wiley, 2007)
3. A.M. Cowley, S.M. Sze, Surface states and barrier height of metal-semiconductor systems. J. Appl. Phys. **36**(10), 3212–3220
4. S.J. Moloi, M. McPherson, Capacitance-voltage behaviour of Schottky diodes fabricated on p-type silicon for radiation-hard detectors. Radiat. Phys. Chem. (2013). https://doi.org/10.1016/j.radphyschem.2012.12.002 (*Article in Press*)
5. A.I. Prokopyev, S.A. Mesheryakov, Restrictions of forward I-V methods for determination of Schottky diode parameters. Measurement **33**(2), 135–144
6. A.I. Prokopyev, S.A. Mesheryakov, Fast extraction of static parameters of Schottky diodes from forward I-V characteristic. Measurement **37**(2), 149–155 (2005)
7. M. Mohammed, Z. Li, J. Cui, T. Chen, Junction investigation of graphene/silicon Schottky diodes. Nanoscale Res. Lett. **2012**(7), 302 (2011). https://doi.org/10.1186/1556-276X-7-302
8. S. Tongay, M. Lemaitre, X. Miao, B. Gila, B.R. Appleton, A.F. Hebard, Rectification at graphene-semiconductor interfaces: zero-gap semiconductor-based diodes. Phys. Rev. X **2**, 011002 (2012). https://doi.org/10.1103/PhysRevX.2.011002
9. LTspice, *High Performance SPICE Simulator*. http://www.linear.com/designtools/software/Accessed Feb. 2013
10. BAT54/A/C/S Rev. E2, *BAT54 Schottky Barrier Diode Product Specification*, Fairchild Semiconductor Corporation (2012)
11. R. Singh, S.K. Arora, R. Tyagi, S.K. Agarwal, D. Kanjilal, Temperature dependence of current-voltage characteristics of Au/n-GaAs epitaxial Schottky diode. Bull. Mater. Sci. **23**(6), 471–474 (2000)

12. R. Sharma, Temperature dependence of I-V characteristics of Au/n-Si Schottky Barrier Diode. J. Electron Dev. **8**(2010), 286–292 (2010)
13. Engauge Digitizer, *Digitizing software*. http://digitizer.sourceforge.net/Accessed Feb. 2013
14. A. Chawanda, W. Mtangi, F.D. Auret, J. Nel, C. Nyamhere, M. Diale, Current-voltage temperature characteristics of Au/n-Ge (1 0 0) Schottky diodes. Physica B: Cond. Matter **407**(10), 1574–1577 (2012)
15. R.O. Ocaya, P.V.C. Luhanga, A fresh look at the semiconductor bandgap using constant current data. Eur. J. Phys. **32**, 1155 (2011). https://doi.org/10.1088/0143-0807/32/5/003
16. R.O. Ocaya, A linear, wide-range absolute temperature thermometer using a novel p-n diode sensing technique. Measurement **46**(4), 1464–1469 (2013). https://doi.org/10.1016/j.measurement.2012.12.008
17. J.R. Taylor, *An Introduction to Error Analysis: The Study of Uncertainties in Physical Measurements*, 2nd ed. (University Science Books, 1996)
18. J.C.G.M 100:2008, *Evaluation of Measurement Data-Guide to the Expression of Uncertainty in Measurement* (2008). (http://www.bipm.org/en/publications/guides/gum.html). Accessed Aug. 2013
19. D.N. Gujarati, *Basic Econometrics*, 4th ed. (McGraw-Hill, 2003)
20. P.M. Gammon, E. Donchev, A. Pérez-Tomás, V.A. Shah, J.S. Pang, P.K. Petrov, M.R. Jennings, C.A. Fisher, P.A. Mawby, D.R. Leadley, N. McN, Alford, A study of temperature-related nonlinearity at the metal-silicon interface. J. Appl. Phys. **112**, 114513 (2012). https://doi.org/10.1063/1.4768718
21. A. Chawanda, K.T. Roro, F.D. Auret, W. Mtangi, C. Nyamhere, J. Nel, L. Leach, Determination of the laterally homogeneous barrier height of palladium Schottky barrier diodes on n-Ge (111). Mater. Sci. Semicond. Proc. **13**(5–6), 371–375 (2010)
22. Keithley, *Model 230 Programmable Voltage Source Instruction Manual*, Keithley Instruments, Inc. Document Number: 230-901-01 Rev. H, Eighth Printing, Cleveland, Ohio, USA, Dec. 2001
23. R.O. Ocaya, PIC and PC integration: Candidates for low cost data acquisition, in *EUROCON 2005—The International Conference on Computer as a Tool II*, Art. no. 1630155, pp. 1144–1147 (2005). https://doi.org/10.1109/EURCON.2005.1630155
24. E.H. Rhoderick, R.H. Williams, *Metal-Semiconductor Contacts* (Clarendon Press, 1988)
25. S. Chand, J. Kumar, Current-voltage characteristics and barrier parameters of Pd_2Si/p-Si(111) Schottky diodes in a wide temperature range. Semicond. Sci. Technol. **10**(12), 1680–1688 (1995)
26. S. Chand, J. Kumar, Evidence for the double distribution of barrier heights in Pd_2Si/n-Si Schottky diodes from I-V-T measurements. Semicond. Sci. Technol. **11**, 1203–1208 (1996)
27. W. Mönch, *Semiconductor Surfaces and Interfaces*, 2nd ed. (Springer, Berlin, 1995)
28. A. Chawanda, J.M. Nel, F.D. Auret, W. Mtangi, C. Nyamhere, M. Diale, L. Leach, Correlation between barrier heights and ideality factors of Ni/n-Ge (100) Schottky Barrier Diodes. J. Korean Phys. Soc. **57**(6), 1970–1975 (2010)
29. H.B. Yao, C.C. Tan, S.L. Liew, C.T. Chua, R. Li, R.T.P. Lee, S.J. Lee, D.Z. Chi, Int. Workshop Junction Technol. Proc. **85** (2005)
30. National Instruments, *White Paper: Tables of Thermoelectric Voltages for All Thermocouple Types*. http://www.ni.com/white-paper/4231/en/#toc1. Accessed Oct. 2013
31. M.K. Hudait, S.B. Krupanidhi, Doping dependence of the barrier height and ideality factor of Au/n-GaAs Schottky diodes at low temperatures. Physica B: Condensed Matter **307**, 125–137 (2007)
32. V. Aubry, F. Meyer, Schottky diodes with high series resistance: limitations of forward I-V methods. J. Appl. Phys. **76**(12), 7973–7984 (1994)

Chapter 6
p-n Diode Parameter Extraction

Abstract One may propose that the actual implementation of a device as a transducer in a sensing application circuit provides strong validation of its extracted inherent parameters. This route is especially true when there are alternative sensors for the purpose of comparing the measurements. This chapter presents a novel application that relies on observations of p-n diode voltage under various constant currents. The measurements are compared with two alternative thermometers. The voltage across a p-n diode is repeatable in this connection. The following discussion demonstrates a novel approach for use in a direct measuring thermometer with a kelvin scale. Importantly, this approach simplifies the complexities often associated with p-n diode thermal sensors by reducing the need for conditioning circuits, eliminating the requirement for delicate voltage references, and removing the need for calibration. In our test application, a microcontroller modulates the current flowing through the diode, switching between two predefined values. The buffered diode voltage is channeled subsequently into the microcontroller's analog-to-digital converter (ADC), enabling straightforward temperature readout on a compatible device with minimal programming requirements. Finally, we assess the prototype's performance against two established standards, thereby providing validation for this innovative approach. The chapter concludes that by implementing a theoretically predicted mode, one can indirectly confirm the validity of the methods of extracting the inherent parameters of the diode.

6.1 Borrowing from p-n Diode Measurements

In the past, many experiments established the voltage-temperature responses of the semiconductor p-n diode with a constant current passing through it. For instance, it is widely accepted that the coefficient of the voltage-temperature response remains constant across a wide temperature range.

© The Author(s), under exclusive license to Springer Nature Switzerland AG 2024
R. Ocaya, *Extraction of Semiconductor Diode Parameters*,
https://doi.org/10.1007/978-3-031-48847-4_6

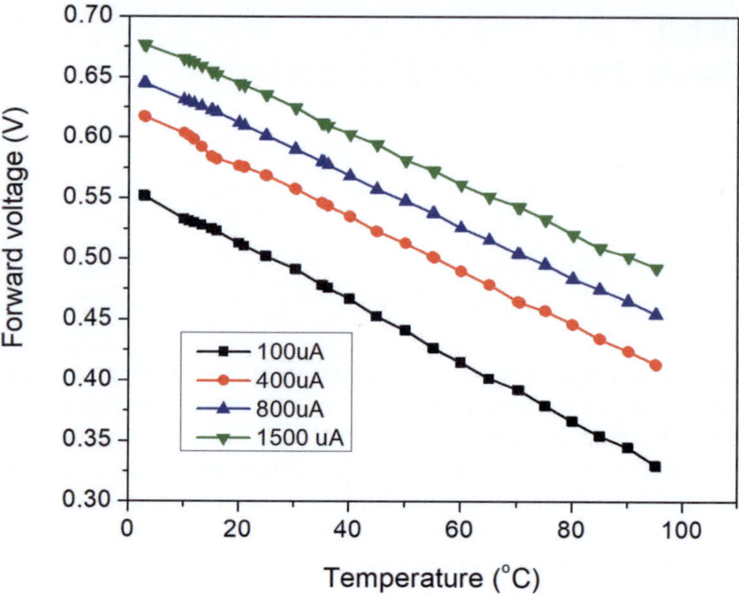

Fig. 6.1 Plot of voltage versus temperature at constant diode current. The gradient of each denotes the temperature coefficient of forward voltage, denoted κ in the text

Ocaya [1], using a purely empirically-based approach, showed that the diode current could be written

$$I = \exp\left(\frac{V - \rho T - \zeta}{\eta T + \epsilon}\right), \tag{6.1}$$

where ρ, ζ, η, and ϵ are constants. For the specific case of the 1N4148 p-n diode, the constants were found to be $\rho = -0.984$ mV/K, $\zeta = 1.25$V, $\eta = 0.155$ mV/K, and $\epsilon = 2.65$ mV.

Figure 6.1 shows the typical results of the experiment where the forward diode voltage is measured at constant diode current as the temperature is varied. If Eq. 6.1 is partially differentiated w.r.t. temperature under the assumption of constant current and then solving for the voltage-temperature rate of change i.e. the temperature coefficient of voltage, gives:

$$\kappa = \left.\frac{\partial V}{\partial T}\right|_{I\ const.} = \rho + \eta \ln(I). \tag{6.2}$$

This result suggests that the temperature coefficient of voltage depends on the logarithm of the actual current passing through the diode [1]. One can verify by substituting the above constants at currents that do not cause diode heating (<1 mA) that, for the 1N4148, $\kappa \approx -2.41$ mV/K. This value is typical for silicon p-n diodes. Figure 6.2 shows the closeness of the model with the actual measured coefficients of the 1N4148 diode.

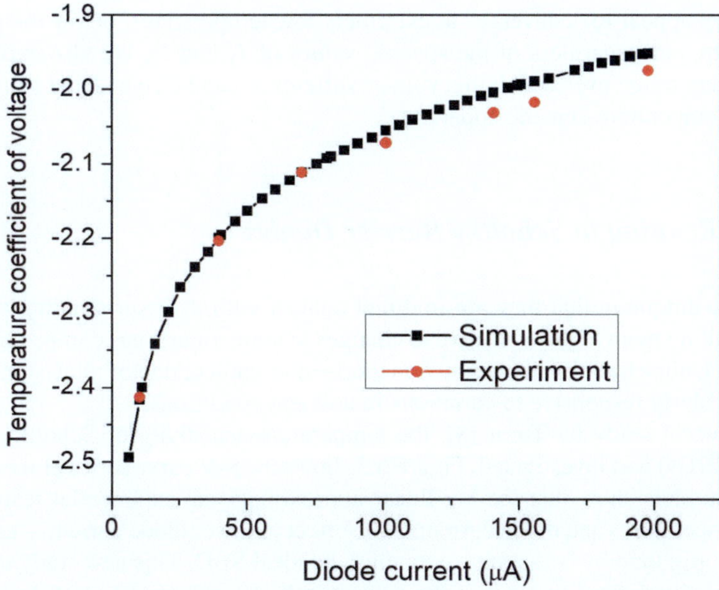

Fig. 6.2 Comparison of the simulated and measured temperature-coefficients of voltage at various diode forward currents

This critical insight forms the basis for designing a diode-based thermometer circuit with a programmable thermal gradient, controlled by the device current. Subsequent investigations [2, 3] delved deeper into the impact of diode material parameters on voltage and current responses. Notably, it provided a more concrete understanding of the significance of factors like the ideality factor and the bandgap, aspects which were routinely assumed by most experimenters [4–7].

A subsequent innovation, described in detail in [2], has proven invaluable for ongoing experiments, including the work presented here. It takes the form of a switchable, dual-constant current source. The ability to switch currents in known ratios allows for the rapid calculation of the bandgap using just three data points, over a narrower temperature range, such as 5 °C. This alternative representation also plays a crucial role in quantifying the effects of bandgap-temperature variations on specific results.

Furthermore, prior research has already established the suitability of certain diode materials for temperature-sensing applications, particularly in terms of the constancy of ideality factors with temperature. Notably, it has been shown that silicon diodes generally exhibit more favorable characteristics in this regard compared to other substrate materials, such as germanium diodes [2].

Here, we present an investigation into the temperature behavior of the observable voltage difference across a silicon diode, initially driven at a constant current (I_1) and subsequently at a different constant current (I_2). Data from earlier experiments [2, 3] suggest that this voltage difference increases with temperature. Intriguingly, when the voltage-temperature responses of any two diodes are extrapolated backward, they appear to "converge" at extremely low temperatures, below the point of liquid nitrogen, regardless of the specific values of I_1 and I_2. We also explore the temperature range over which this voltage difference can be considered suitable for use in a temperature-sensing apparatus.

6.1.1   Relating to Schottky Barrier Diodes

SBDs are unique in that they are in direct contact with their surrounding environment, making them highly sensitive to changes in temperature, gas composition, and pressure. Unlike traditional p-n junction diodes and semiconductor transistors, SBDs are particularly responsive to variations in ambient conditions.

In a recent study by Turut [8], the temperature sensitivity of Schottky barrier diodes (SBDs) was investigated. Figure 6.3 shows the I–V curve series of the studied SBD. The plots show that the log-linear approximates an exponential response at lower temperatures and diode currents. The effect of lower diode current is to reduce the $I R_s$ term, thereby better approximating the ideal SBD. This new study involved numerical simulations to analyze the current-voltage characteristics of Au/n-GaAs SBDs concerning temperature sensitivity. The findings of the simulation study indicated that the thermal sensitivity of SBDs increased as the current level decreased, assuming the same SC area size (Fig. 6.4).

Conversely, when the SC area size was reduced, the thermal sensitivity decreased for the same current level. Their results suggest that SBDs with larger SC areas should be operated at lower current levels, especially in cryogenic temperature ranges (Fig. 6.5).

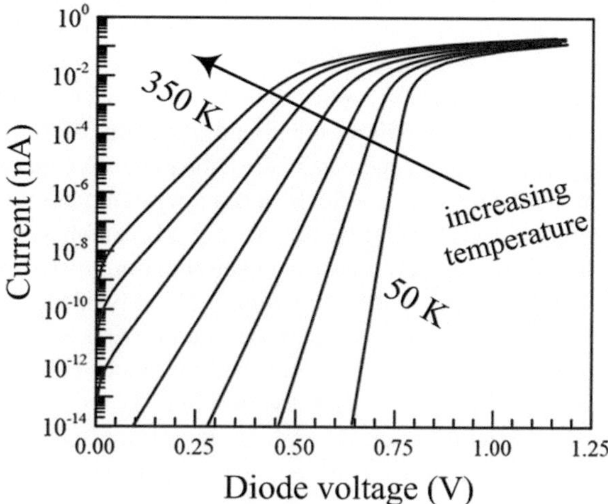

Fig. 6.3 Simulated forward current versus voltage and temperature response of an Au/n-GaAs SBD of 1 mm diameter. The curves are simulated in temperature steps of 50 K. Adapted from Turut [8]

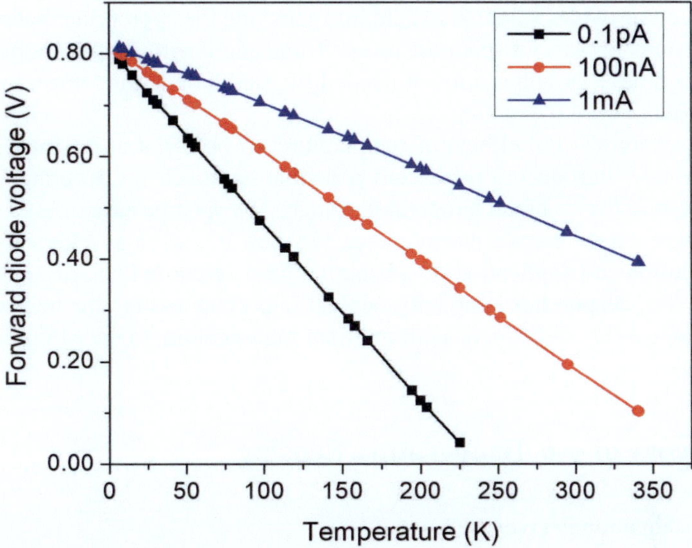

Fig. 6.4   Forward diode voltage versus temperature response at constant current of an Au/n-GaAs SBD of 1 mm diameter. Adapted from Turut [8]

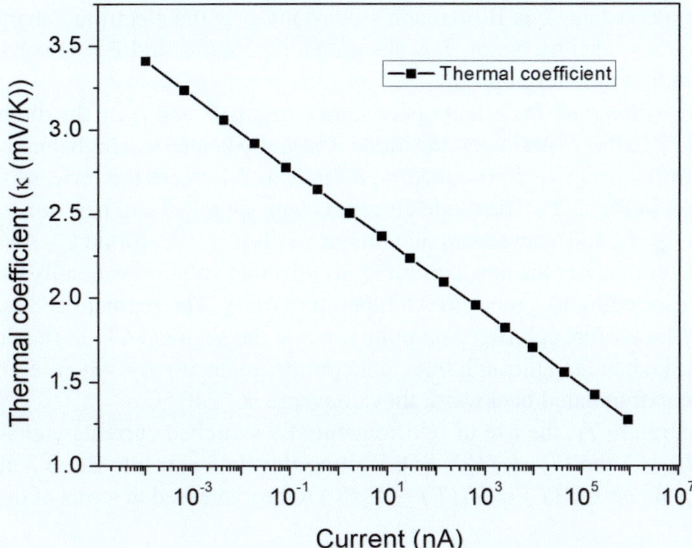

Fig. 6.5   Observed thermal coefficient of forward diode voltage for a Au/n-GaAs SBD of 1 mm diameter. Adapted from Turut [8]

This recent provides valuable insight into selecting the appropriate Schottky contact dimensions for various scientific research and engineering applications. Perhaps for the first time, it provides a much-needed focused look at the temperature sensitivity of SBDs.

Having examined the effects of temperature on both p-n and Schottky barrier diodes, we now introduce a novel and plausible approach for accurately measuring temperature over a broad temperature range. The specific case offers a potential application in a Kelvin-scale thermometer. We view this work as a logical yet innovative evolution and application of substantial prior research focused on the simple p-n diode. Yet, despite this simplicity, several important lessons can be learned that are applicable either directly, or with minimal modification, to the MS diode.

6.2   Theory of p-n Temperature Sensing

Shockley's equation is given by Eq. 6.3:

$$I \approx I_n \exp\left(-\frac{E_g}{nkT}\right) \exp\left(\frac{qV}{nkT}\right). \tag{6.3}$$

Here, I_n is a constant, k is Boltzmann's constant, q is the electronic charge, n is a diode-dependent ideality factor, T is absolute temperature, and E_g is the bandgap of the semiconductor material used.

Figure 5.1 illustrates the effect of constant currents I_1 and I_2 on the diode voltage as it cools [2]. In this experiment, the diode is initially heated to a high temperature T_0 under forward current I_1. Subsequently, it cools to T_1, where the forward voltage is V_{F1} (point B in Fig. 5.1). The diode current is then switched to a new constant value I_2, where $I_2 \leq I_1$, and a new terminal voltage V_{F2} is observed (point C). As the diode continues to cool under the new current I_2, its terminal voltage eventually reaches V_F again, corresponding to a measured temperature of T_2. The segment BC establishes the ideality factor through two data points, while the segment CD in relation to BC establishes the bandgap through three data points. Interestingly, when segments AB and CD are extrapolated backward, they converge at –240 °C.

At temperature T_1, the use of two sequentially switched currents yields the data pair: (T_1, I_1, V_1) and (T_1, I_2, V_2). Substituting this data pair into Eq. 6.3 allows the difference voltage $\Delta V(T) = V_1(T) - V_2(T)$ to be expressed in terms of the ideality factor as:

$$\Delta V(T_1) = \left(\frac{nk}{q} \ln \frac{I_1}{I_2}\right) T_1. \tag{6.4}$$

Suppose that the two currents are in a known ratio a, such that:

$$I_2 = aI_1, \quad \text{or} \quad m = \ln\left(\frac{I_1}{I_2}\right), \tag{6.5}$$

where $0 < a < 1$. If one defines a new temperature T^* s.t.

$$\frac{1}{T^*} = \frac{1}{T_2} - \frac{1}{T_1}, \tag{6.6}$$

then the bandgap can be found:

$$E_g = V_F - \frac{nkT^*}{q} \ln a \quad \text{(in electron-volts, eV)}. \tag{6.7}$$

Furthermore, rearranging Eq. 6.4 for the ideality factor n at $T = T_1$ and substituting into Eq. 6.7 yields an alternative bandgap equation:

$$E_g = V_F + \frac{T_2}{T_1 - T_2} \Delta V(T_1). \tag{6.8}$$

Thus, the bandgap is determinable by measuring the voltage change using only three points along two different voltage-temperature paths provided the current ratio remains unchanged. The uncertainty in the bandgap can be estimated using:

$$\Delta E_g \approx \pm \frac{2\alpha T}{\beta + T} \Delta T. \tag{6.9}$$

For silicon $E_g(0) = 1.17$ eV, $\alpha = 0.473$ milli-eV/K, $\beta = 636$ K [4, 9], and ΔT is the temperature uncertainty. Equation 6.9, with $T = T_1$, allows for determining the effect of bandgap variation with temperature on $\Delta V(T)$. Impurities [10] and doping density [11] also affect bandgap but are beyond our control with commercial diodes. The typical ideality factor for the 1N4148 silicon diode is circa 1.9 [2]. In Eq. 6.7, the terms n, q/kT, and $m = \ln(I_1/I_2)$ being constant permit writing in terms of a new constant b:

$$\Delta V(T) = bT, \tag{6.10}$$

where $b = (nk/q)m$ in J/K.C or V/K (volts per kelvin). Therefore, the voltage difference at a given temperature T is directly proportional to the temperature provided that the currents are switched in ratio at the measurement temperature.

The switch-over (T_1, along segment BC) must be faster than the rate of temperature change. Equation 6.10 implies that higher measurement accuracy is possible if the voltage difference $\Delta V(T)$ is large. However, a large ratio I_1/I_2, especially if achieved by making I_1 much larger than I_2, could lead to greater self-heating with the overall result of lowered temperature measurement accuracy.

6.2.1   Simulation Approach

The LTSpice simulation circuit, which consists of a single current source in series with the diode, is illustrated in Fig. 6.6.

The circuit design did not contain any elements affected by temperature changes within the specified sweep range, except for the diode. LTSpice is unable to simulate temperatures lower than –268 °C. The LTSpice statement used for the simulation is as follows:

```
.dc temp -260 200 1 I1 118u 200u 82u
```

This statement initiates the simulation with two currents: $118\,\mu A$ and $200\,\mu A$, and the temperature is swept from –260 °C to 200 °C in 1 °C increments.

6.2.2   Experimental Characterization

The diode characterization experiment involved cooling a p-n diode in a constantly stirred oil bath previously heated to about 100 °C. The use of oil eliminates the need for electrical insulation of the diode while improving thermal contact with the bath, whose temperature was monitored constantly by a Major Tech MT630 digital thermometer. The diode voltage was measured using a Yokogawa 7533-05 digital multimeter. The switchable constant current described in [2] was configured for the six parametric ratios (m) in Table 6.1.

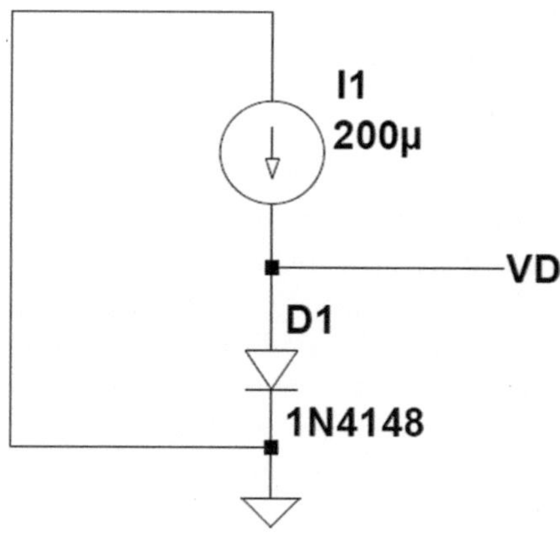

Fig. 6.6 Simulation circuit used to investigate ΔV versus temperature behavior of the diode

6.3   Results and Analysis

6.3.1   Simulation Results

Figure 6.7 depicts the impact of the variation of *m* on the voltage-temperature response of the diode. A lower value of *m* implies that the current I_1 is generally lower. Figure 6.7 suggests that at the upper-temperature limit, the diode response becomes non-linear earlier at the two extremes of *m* (at 120 °C as compared to 160 °C) as *m* increases, although the difference voltage (ΔV) is about 4.1 times higher at 134.8 mV, which should make its measurement easier.

There is a trade-off between the desired linear range and the output level during the subsequent design of the sensor design. A simple non-inverting amplifier stage boosts the output level in any case. Therefore, the lower value of *m* is a better choice. In earlier work [1–3], the reported results were solely of experimental origin, but these simulation results are sufficiently supportive of the experiments.

6.3.2   Diode Thermometer Operation

The experiments involved taking measurements as the diode cooled from about 100 °C to about 0 °C. In the experiment, the currents were varied over a wide range to permit a wider spread of the parameter *m*, as shown in Table 6.1. The actual currents used were 100, 400, 800, and 1500 μA.

Figure 6.8 graphically shows the effect of varying *m* in the experiment. Although not shown, when extrapolated backward, the curves intersect around 33.4 K or –240 °C. The gradients of the curves obtained by linear fitting are given in

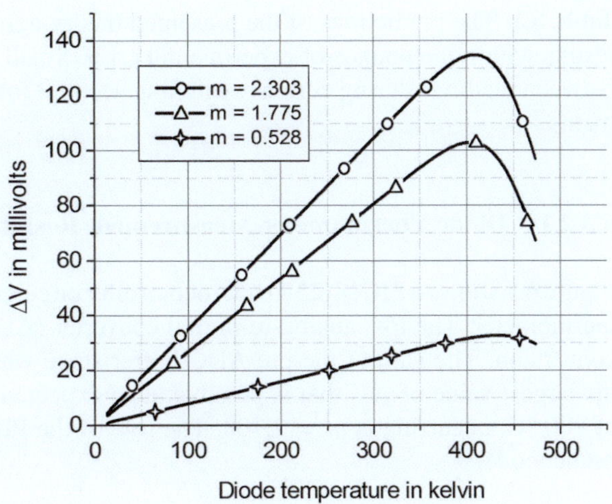

Fig. 6.7 Simulated effect of varying parameter *m* on the diode's temperature response

Table 6.1 Summary of theoretically and experimentally determined values of b in volts per Kelvin

Theoretical		Experimental (linear fit)	
$m = \ln(I_1/I_2)$	b (V/K)	b (V/K)	% error in b
0.629	1.029E–04	8.490E–05	−21.2
0.693	1.134E–04	9.120E–05	−24.3
1.322	2.163E–04	1.952E–04	−10.8
1.386	2.269E–04	2.374E–04	4.4
2.079	3.403E–04	3.283E–04	−3.7
2.708	4.432E–04	4.168E–04	−6.3

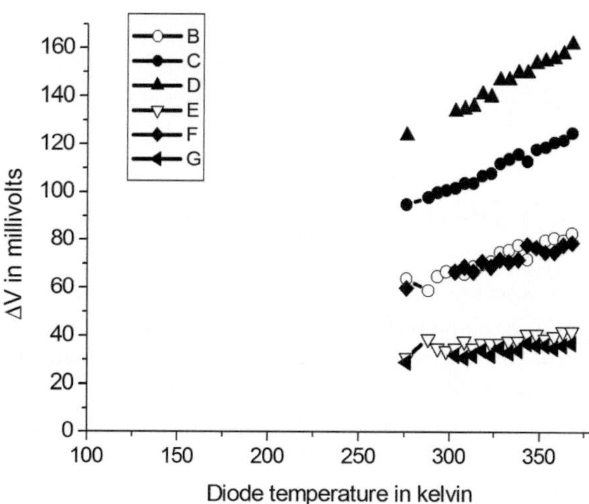

Fig. 6.8 Experimental effect of varying parameter m on the diode's temperature response

Table 6.1. The predictions of the presented theory agree well with the experimental results within the bounds of experimental error. Small drifts in the bath temperature between readings during current-switching account for the observed slight disagreements.

6.3.2.1   Diode Thermometer Measurement Results

Figure 6.9 shows a PIC18f2550 microcontroller circuit used as a test bed for the diode thermometer. The PIC analog-to-digital converter (ADC) was configured for 10-bit conversion. The gain of the pre-ADC buffer stage was set at 3.055 by considering the largest value of ΔV that is possible at an expected upper operating temperature of 200 °C, occurring at $m = 2.708$. The role of the PIC is defined by the following pseudocode:

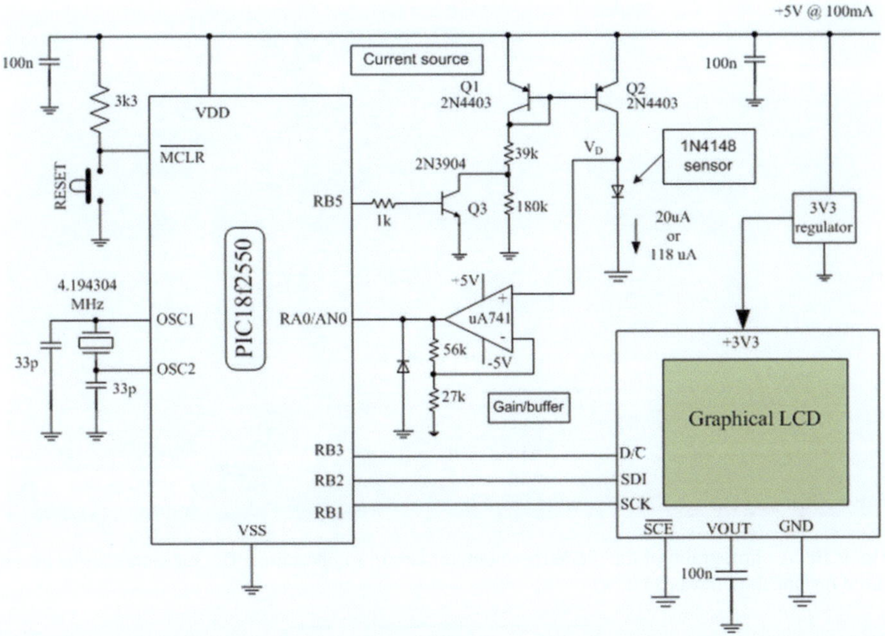

Fig. 6.9 Complete schematic diagram of the thermometer. The diode at AN0 protects the PIC by clamping negative OPAMP outputs to –0.7 V

1: Define parameters $b = 2.6 \times 10^{-4}$ and $A_F = 3.055$
2: Switch diode current to I_1, then read $A_F V_{F1}$
3: Switch diode current to I_2, then read $A_F V_{F2}$
4: Calculate the true diode voltage difference in Fig. 5.1: $\Delta V = (V_{F1} - V_{F2})/A_F$
5: Calculate kelvin temperature using $T = \Delta V / b$
6: Display the temperature on the readout device in the correct units (K, °C or °CF)
7: Wait the update period e.g. one second, then repeat indefinitely from step 1.

In this test, the mikroC compiler [12] was used to code the microcontroller in the thermometer. Figure 6.11 shows the results of an experiment to compare the performance of the diode thermometer versus the MajorTech MT630 digital thermometer against a standard laboratory alcohol-in-glass thermometer.

This experiment does not require sophisticated apparatus. A water-ice bath heated from 0 °C to around 100 °C suffices. Figure 6.11 shows the performance of the diode thermometer against the commercial MT630 thermocouple thermometer.

The experimental results show that the agreement between the alcohol reference thermometer and diode thermometer (at 0.1 °C on average) is generally better than that with the MT630 (at 1 °C on average). There are a number of possible reasons for this difference in response. Firstly, the bulkiness of the MT630 probe likely introduces a lag in its response to heating. Additionally, convective effects within the water bath could have added to the lag observed in both thermometers, which cannot physically

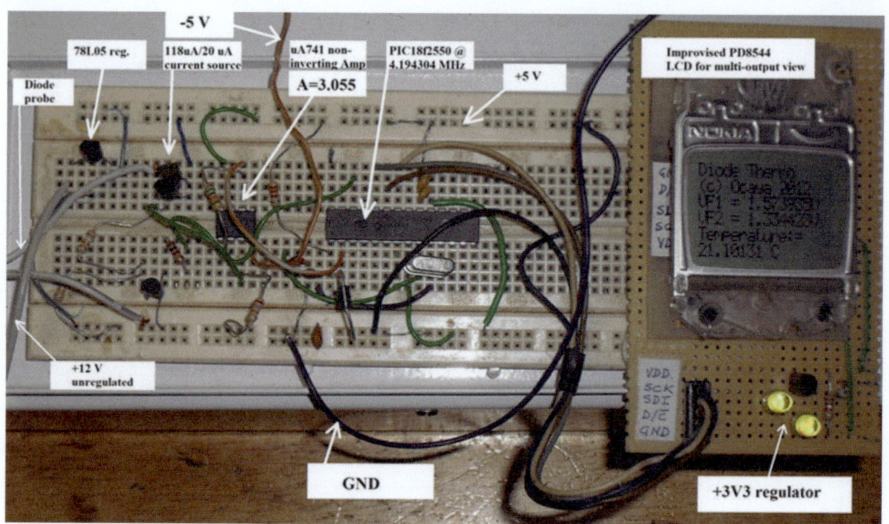

Fig. 6.10 A photograph of the diode thermometer circuit in operation. The measured temperature at the instant shown was 21.1 °C

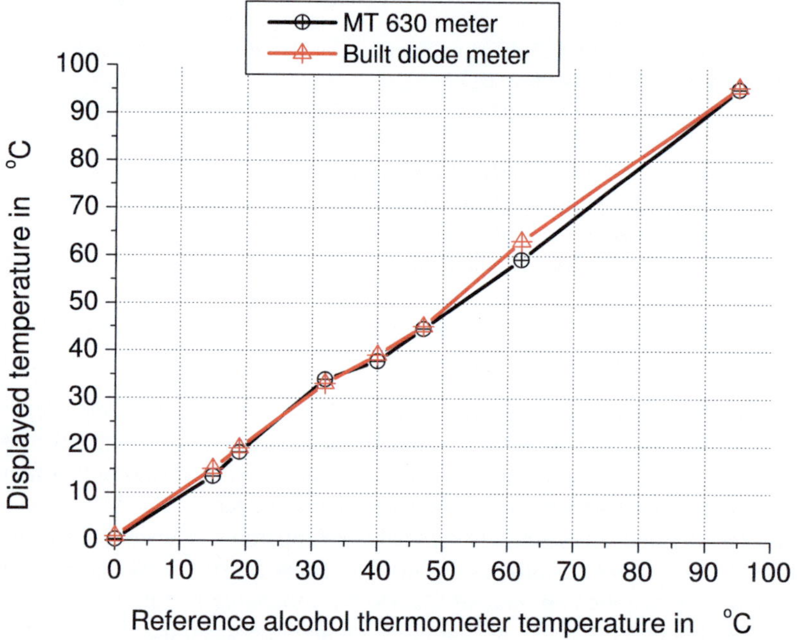

Fig. 6.11 Performance comparison of the diode thermometer against a MajorTech MT630 digital type and laboratory alcohol in bulb type

be at the same point and simultaneously experience the same temperature. Secondly, drifts or calibration errors in the MT630 thermocouple compensation circuitry may have affected its overall accuracy. However, within the limits of these errors, the experimental results are in good agreement (Fig. 6.10).

6.4 Confirming Parameter Through Application

The p-n diode has been studied extensively for decades for its potential as a thermometer. Its negative thermal voltage coefficient is relatively constant when the diode current is constant. However, traditional circuits for obtaining direct temperature readings from diode sensors typically require complex conditioning. They need operational amplifier modules that implement gain, subtractors, voltage references, and calibration. The above discussion presented a novel approach to leverage the voltage difference across the diode at the same temperature to create a linear and wide-range diode thermometer. The theoretical analysis and simulations culminated in a test circuit validation that confirmed the theory.

The findings indicate that this novel thermometer operates on the kelvin scale due to the constant b. The experimental results compare favorably with a standard alcohol-in-glass thermometer and a commercial thermometer, the MajorTech MT630, proving the approach is feasible. It is possible to refine the circuit in many ways. For one, the stability of the current source is linked directly to the reliability of the measurements. The precise values of the currents are relatively unimportant since they are used in ratio. The circuit's applicability is expected to span a wide temperature range, from approximately –200 °C (cryogenic) to about 200 °C, though precise testing across this entire range remains a topic for future investigation.

6.4.1 Recent Comparisons

In a similar recent study, Gumus and Aydogan [13] investigated Ni/p–GO@Fe$_3$O$_4$/p–Si(P–P) and Ni/p–GO@Fe$_3$O$_4$/n–Si(P–N) heterojunctions (HJs) constructed under the same conditions. They conducted current-voltage (I–V) measurements across a wide temperature range and applied thermionic emission theory to analyze the temperature-dependent I–V characteristics of the devices. The main parameters of the HJs, including the ideality factor (n), barrier height (Φ), rectification ratio (RR), and turn-on voltage (V_0), were examined as functions of temperature. The researchers observed that all these parameters exhibited temperature-dependent variations.

Specifically, they noted that RR and V_0 values for both devices decreased as the temperature increased, while Φ increased. This behavior is attributed to the heterogeneity between p-GO@Fe$_3$O$_4$ and Si.

Gumus and Aydogan also calculated the thermal sensitivity (S) and activation energy for both devices. They found that S exhibited a linear decrease with increasing current. Additionally, they determined the highest sensitivity to be 1.35 mV/K with a corresponding activation energy of 0.79 eV for the P–P HJ. For the P–N HJ, these values were calculated as 0.97 mV/K and 0.66 eV, respectively. This study provides valuable insights into the temperature-dependent behavior of these heterojunctions and their potential applications in thermal sensing.

6.5 Remarks

This chapter introduced a novel application that validates the extraction of inherent parameters in semiconductor devices by employing them as transducers in sensing circuits. This approach directly validates parameter extraction by using the device in a realistic setting in such a way that the results can be compared with standardized measuring instruments. Specifically, the approach involves comparing measurements from a p-n diode voltage, obtained under different constant currents, with those from two alternative thermometers. The p-n diode voltage measurements are found to be repeatable, leading to the development of a direct measuring thermometer with a Kelvin scale.

References

1. R.O. Ocaya, An experiment to profile the voltage, current and temperature behaviour of a P-N diode. Euro. J. Phys. **27**, 625 (2006). https://doi.org/10.1088/0143-0807/27/3/015
2. R.O. Ocaya, P.V.C. Luhanga, A fresh look at the semiconductor bandgap using constant current data. Euro. J. Phys. **32**, 1155 (2011). https://doi.org/10.1088/0143-0807/32/5/003
3. R.O. Ocaya, F.B. Dejene, Estimating p-n diode bulk parameters, bandgap energy and absolute zero by a simple experiment. Euro. J. Phys. **28**, 85 (2007). https://doi.org/10.1088/0143-0807/28/1/009
4. J.W. Precker, M.A da Silva, Am. J. Phys. **70**(11) (2002)
5. P.J. Collings, Simple measurement of the band gap in silicon and germanium. Am. J. Phys. **48**, 197–199 (1980)
6. C.W. Fischer, Elementary technique to measure the energy band gap and diffusion potential of pn junctions. Am. J. Phys. **50**, 1103–1105 (1982)
7. R. Mukaro, B.M. Taele, D. Tinarwo, Euro. J. Phys. **27**, 531 (2006)
8. A. Turut, Theoretical approach to thermal sensitivity capability of metal-semiconductor diodes with different Schottky contact area. J. Vacuum Sci. Technol. B **41**(6) (2023). https://doi.org/10.1116/6.0002976
9. W. Bludau, A. Onton, W. Heinke, Temperature dependence of the band gap of silicon. J. Appl. Phys. **45**, AIP, 1846–1848 (1974)
10. R.A. Abram, G.N. Childs, P.A. Saunderson, Band gap narrowing due to many-body effects in silicon and gallium arsenide. J. Appl. Phys. **45**, 1846–1848 (1974)
11. Y. Pan, M. Kleefstra, Minority carrier transport equations in heavily doped silicon including band tail effects at thermal equilibrium. JPC **17**, 6105–6125 (1984)
12. MikroElektronika, *Mikroc Pro for PIC Microcontrollers*. Last accessed August 2023
13. I. Gumus, S. Aydogan, Thermal sensing capability of metal/composite-semiconductor framework device with the low barrier double Gaussian over wide temperature range. Sens. Actuators A: Phys. **332**, 113–117 (2021). https://doi.org/10.1016/j.sna.2021.113117

Chapter 7
A Novel Approach Based on Symmetry

Abstract This chapter introduces a concise I–V method for extracting key parameters of metal-semiconductor diodes, such as series resistance, ideality factor, barrier height, and built-in potential. It achieves this by reformulating the TE equation into an ordinary differential equation, revealing an unexpected symmetry that eliminates the series resistance term. Although the method's derivation may appear complex, its practical application is straightforward. The results obtained using this method align strongly with established trends in the literature compared to other electrical methods. Notably, series resistance calculations from this method exhibit rare agreement with the C–V method compared to the Cheung-Cheung method, accurately reproducing additional parameters like photo-conductive response and interface state density.

7.1 Introduction

Quantifying the parameters Φ, R_s, and n is a fundamental endeavor in semiconductor materials research. A spectrum of techniques, encompassing optical spectrometry, impedance analysis, and electrical measurements, has been developed for this purpose, each with its distinctive advantages and constraints. Amid these methods, electrical approaches stand out for their versatility, albeit accompanied by varying levels of intricacy. The intricacies are rooted in their unique methodologies and the potential interdependencies among the parameters of interest [1, 2]. Adding to the challenge, these intricacies are further amplified by the cyclic nature of current expressions.

Within the framework of thermionic emission (TE), the extraction of Schottky diode parameters poses a formidable challenge due to these intricate parameter connections. It's noteworthy that most Schottky diodes, in reality, manifest as metal-semiconductor-metal (MSM) devices, resembling two asymmetric back-to-back diodes [3] as shown in Fig. 7.1. Consequently, achieving a purely Schottky-type behavior, where one junction rectifies while the other exhibits purely Ohmic behavior, remains an elusive goal. Nevertheless, an interesting concept has emerged, suggesting that minor variations in barrier heights could be disregarded, ultimately leading to the development of methodologies with a central focus on R_s [4, 5].

  119
R. Ocaya, *Extraction of Semiconductor Diode Parameters*,
https://doi.org/10.1007/978-3-031-48847-4_7

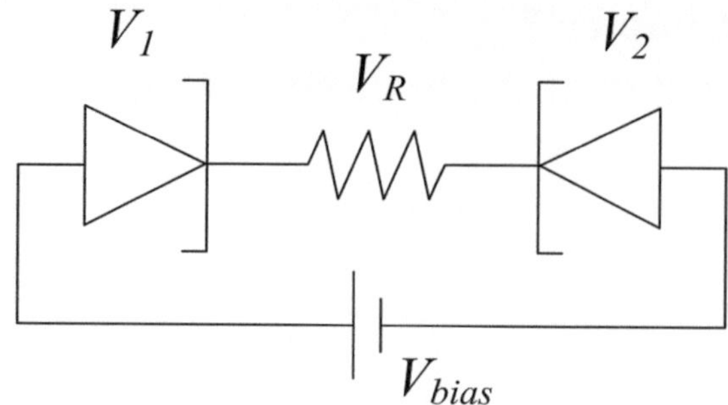

Fig. 7.1 Back-to-back configuration of some Schottky diodes created under some fabrication conditions

These methodologies follow an algorithmic approach that effectively suits empirical datasets and proves applicable to a broad spectrum of MS devices and MSM diodes.

7.2   Unification by Symmetry

The complexity of the MS diode equation accounts for the proliferation of the electrical parameter extraction methods. Sensitivity to applied bias and the parametric interdependence of the parameters themselves poses additional challenges. The different methods thus lead to considerable variance in the reported barrier height for a given device. This variation has been addressed by many workers in the past through a Gaussian distribution that collectively encapsulates them as barrier height inhomogeneity [6]. A peculiarity evident in the proliferation of methods is the often conflicting trends in the behavior of a variable. That is, whereas one method might show that R_s increases with bias, another might suggest the opposite [15]. Therefore, a unifying approach is needed to consistently calculate the parameters, with low variance, thus imparting a desirable repeatability to the method.

In this chapter, we discuss a newly discovered method that addresses the challenge of Φ, R_s, and n extraction by first putting the MS-device electrical characteristics into the unsuspected framework of group theory [7]. Ultimately, the method leads to an extremely simple representation of the TE equation that does not contain R_s and may thus be named the R_s-compensated TE method [16, 17]. Notably, the method does not presuppose diode structure but strictly adheres to the TE model. It reformulates the TE equation into an ODE, revealing an unexpected symmetry that eliminates the series resistance term. Although the method's derivation may seem intricate, its practical application is straightforward. The results it obtains closely align with established trends in the literature, especially when compared to other electrical methods, but outdo them over a wide range of applied biases. Notably, it calculates parameter values that agree remarkably well with the C-V method compared to the other existing electrical methods.

7.3   The Definition of Symmetry

Figure 7.2 illustrates the essence of the symmetry concept.

In the figure, the square labeled '1' is subjected to several different mathematical operations e.g. translations, scaling, and rotations. Its basic square shape is not changed after the transformation. In contrast, the square labeled 'A' undergoes a series of operations that completely deform it into a trapezium, clearly a non-square. The numerical objects and their operations in the figure are well-defined and form a group, whereas the alphabetical objects and their transformation operations do not. Here, we will present only the essential ideas of symmetry and avoid a deeper and more abstract theory of symmetry groups. But briefly, a group exists if

1. there is a well-defined collection of objects forming a set,
2. there is a well-defined action operator, and
3. there exists an identity operator.

These conditions imply *closure* of the group, i.e. the action of the operator generates new objects that *also* belong in the set. The collection of objects and its operators form a *field*.

The existence of symmetry groups is easy to ascertain visually for geometrical objects. This is certainly not to say that the concept applies only to geometry. Many systems have mathematical functions that form symmetry groups. Often, a clue to the possible existence of a group is the appearance of the function forms when plotted. The new method described in this chapter originated from the observation of similarities in the I–V characteristics of Schottky diodes. Figure 7.3 illustrates the idea using two typical I–V curves, Curve 1 and Curve 2.

In the figure, one might suspect the existence of some kind of symmetry. This is because first, at a given temperature, the serialized semiconductor curves are similar such that Curve 1 ends up at Curve 2 by just a sequence of translations and scalings. Second, both curves clearly originate from the same device and, therefore, form

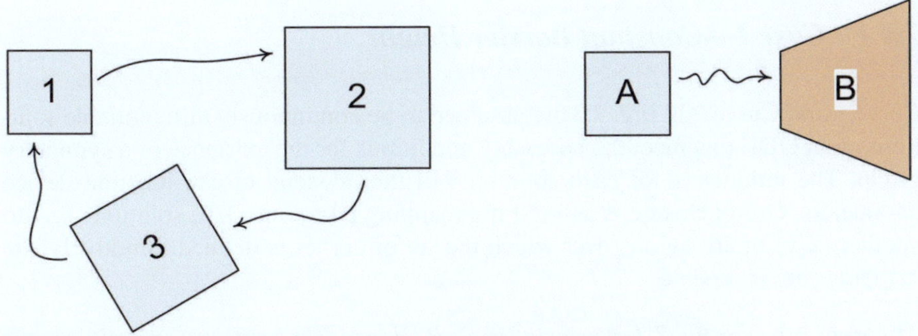

Fig. 7.2   Operations (1 → 2 →3 → 1) are symmetries, but (A→B) is not

members of some set. For semiconductors, a similar curve series is observed when
the device temperature is varied over the same range of bias.

7.4   The Search for I–V Symmetry

Equation 2.26 can be written in a brief general form:

$$y = (1/s)e^{[-b\Phi(x)+c(x-yr)]}\left\{1 - e^{-b(x-yr)}\right\}, \tag{7.1}$$

where $x = V$, $y = I$, $b = q/kT$, $c = b/n$, $r = R_s$, and $(1/s) = AA^*T^2$. Existing
methods estimate Φ and n based on simplifying approximations. A commonly used
assumption is that $qV \geq 3kT$ such that n, R_s are estimated through $\ln(I)$ and $\ln(V)$
functions, which limits the method to the forward low-bias region.

   In general, the barrier height depends on applied bias i.e. $\Phi = \Phi(V)$, leading to
many simplifying assumptions in the literature [8–13]. If Φ is assumed to be constant
then Eq. 7.1 takes the simpler form:

$$y = (1/p)e^{c(x-yr)}\left\{1 - e^{-b(x-yr)}\right\}, \tag{7.2}$$

where $s = p\exp(-b\Phi_0)$, where $\Phi = \Phi_0$ is the zero-bias barrier height. This parameter
is itself of interest. Therefore, two broad treatment cases are possible depending on
whether or not Φ depends on bias. That is

- Case I, given by Eq. 7.2 where $\{p, b, c, r\}$ are constants
- Case II, given by Eq. 7.1.

Both these cases are addressed below.

7.4.1   Case I—Constant Barrier Height

Curve 1 and Curve 2 in Fig. 7.3 are assumed to be continuously differentiable solutions of an ODE that meet the necessary conditions for the existence of a symmetry
group. The uniqueness of each solution will then depend on the inherent device
parameters. Group closure is assured if a mapping takes one ODE solution (x,y) to
another $(\hat{x},\hat{y})$ in all regions over which the set of curves is defined. Intuitively, the
mappings are reversible.

Theorem 7.1 *For Eq. 7.1, a translational, R_s-dependent symmetry reversibly maps
all the device I–V characteristics.*

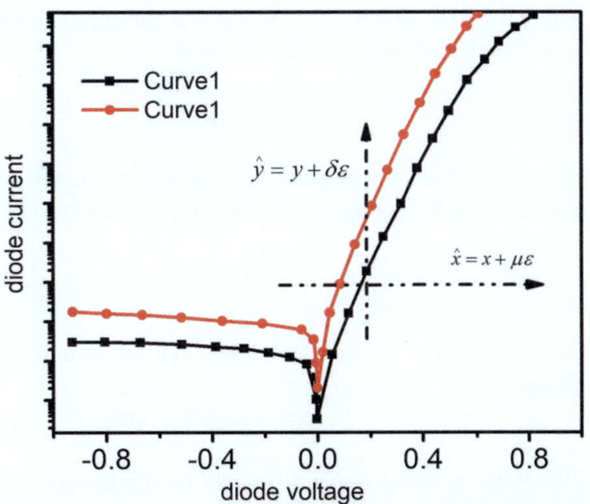

Fig. 7.3 Typical parametric plot of current-voltage characteristics of a Schottky diode with series resistance. The translational symmetry is apparent from Curve 1 to Curve 2 along both x and y axes. There is possibly a scaling symmetry along y

Proof The idea is to construct an ODE from the known solution (Eq. 7.1), and then to show that the ODE structure is preserved under the action of translations. Differentiating it w.r.t x gives

$$py' = be^{c(x-yr)}e^{-b(x-yr)} - bre^{c(x-yr)}e^{-b(x-yr)}y'$$
$$+ c\left\{e^{c(x-yr)} - e^{c(x-yr)}e^{-b(x-yr)}\right\}(1 - ry'), \tag{7.3}$$

where $y' = dy/dx$. Rearranging Eq. 7.3 gives the first-order ODE:

$$y' = \frac{\beta(x, y)}{p + r\beta(x, y)} = \omega(x, y), \tag{7.4}$$

where $\beta(x, y)$

$$\beta = \beta(x, y) = \left[(b - c)e^{-b(x-yr)} + c\right]e^{c(x-yr)}. \tag{7.5}$$

Equation 7.4 is the diode conductance at the point (x,y). It can be estimated accurately from an empirical data plot, providing another way to exploit Eq. 7.4. Clearly, it depends on r, on absolute diode temperature (through b and c), and on $p = p(A, A^*, b, \Phi_0)$. Devices with $r = 0$ are the special case, so that $y' = \beta/p$, where β is itself a simpler expression. However, such devices do not really exist, leading us to focus on the general case with $r \neq 0$.

Overall, the interest is to find the non-trivial symmetries i.e. those that do not map the solution (x,y) onto itself. That is, the action $(x, y) \mapsto (\hat{x}, \hat{y})$ should transform Eq. 7.4 to

$$\hat{y}' = \frac{\hat{\beta}}{p + r\hat{\beta}}, \tag{7.6}$$

where $\hat{\beta} = \hat{\beta}(\hat{x}, \hat{y}) = \beta(\hat{x}, \hat{y})$. In Fig. 7.3, the apparent translational symmetry suggests candidate transformations of the form

$$\hat{x} = x + \mu\varepsilon, \quad \text{and} \quad \hat{y} = y + \delta\varepsilon, \tag{7.7}$$

where ε is the infinitesimal variational parameter and μ and δ are determinable constants. Substituting Eq. 7.7 into Eq. 7.5 leads to

$$\hat{\beta} = \left[(b - c)e^{-b(\hat{x}-\hat{y}r)}e^{b(\mu\varepsilon-\delta\varepsilon r)} + c\right]e^{c(\hat{x}-\hat{y}r)}e^{c(\mu\varepsilon-\delta\varepsilon r)}. \tag{7.8}$$

Comparing Eqs. 7.5 and 7.8 shows that the structure of β is preserved iff $\mu = \delta r$. Hence

$$\hat{x} = x + \delta r\varepsilon, \quad \text{and} \quad \hat{y} = y + \delta\varepsilon. \tag{7.9}$$

Therefore, μ and δ depend linearly on diode series resistance. Also, the condition $\hat{y}' = y'$ still holds by differentiation w.r.t x in both spaces. Finally, Eq. 7.6 arises by substituting $x = (\hat{x} - \delta r\varepsilon)$ and $y = (\hat{y} - \delta\varepsilon)$ into Eq. 7.4.

Thus the only requirement of the symmetry method is the existence of a well-defined group whose membership is the MS device characteristics and the allowed transformations. With reference to Fig. 7.3, this requires that the device resistance r be well-defined for Curve 1 and Curve 2 near ε to validate the Eqs. 7.4 and 7.6. This is true for the empirical data set. Practical measurements on the same diode have characteristics that differ only due to a variation in another parameter, such as T, n, illumination, etc. Thus we could consider the general case when Curve 1 $\neq$ Curve 2, and the specific case when Curve 1 $=$ Curve 2. These possibilities are further discussed later in this chapter. We start at the necessary coordinate transformation, which means moving from the original I–V characteristic to a resistance-compensated I–V characteristic. The small variational parameter above ($\varepsilon \to 0$) allows near-identity transformations on the (x, y) plane:

$$\hat{x} = x + \varepsilon\xi(x, y) + \mathcal{O}(\varepsilon^2)$$
$$\hat{y} = y + \varepsilon\eta(x, y) + \mathcal{O}(\varepsilon^2), \tag{7.10}$$

which, in effect, implements a linearization in terms of the tangent vectors:

$$\xi(x, y) = \left.\frac{d\hat{x}}{d\varepsilon}\right|_{\varepsilon=0}, \quad \text{and} \quad \eta(x, y) = \left.\frac{d\hat{y}}{d\varepsilon}\right|_{\varepsilon=0}. \tag{7.11}$$

The parameter ε is analogous to time (t) in Newtonian mechanics s.t. x and y are the analogies of displacement and velocity, respectively. At $t = 0$, the vector field (ξ, η) would then denote the instantaneous velocity and acceleration, respectively. However, if t is very small but non-zero i.e. $t \to 0$, then (ξ, η) would be the averaged displacement, velocity, and acceleration in the vicinity of t. More on this point in a later part of the discussion.

With symmetry, the variation of ε describes all possible orbits s.t. $(x, y) \mapsto (\hat{x}, \hat{y})$. Equation 7.9 implies that $\xi(x, y)=\delta r$, and $\eta(x, y)=\delta$. The mapping can be expressed instead in terms of some characteristic function $Q=Q(x, y, y')$ s.t.

$$\hat{x} = x, \qquad \hat{y} = y + \varepsilon Q(x, y, y') + \mathcal{O}(\varepsilon^2), \quad \text{and}$$

$$Q(x, y, y') = \eta(x, y) - \xi(x, y)y'. \tag{7.12}$$

The trivial mapping $(\hat{x}, \hat{y}) = (x, y)$ results when $Q = 0$. The trivial mapping is not a particularly useful one since it gives the same solution point and is, therefore, to be excluded generally.

The ODE in the (x, y) coordinate system is transformed into a canonical system that is hopefully easier to solve. The motivation is that the ODE becomes invariant along one axis, thus reducing the solution to just quadrature i.e. integration. For instance, defining a new coordinate system $(f(x, y), g(x, y))$ does not violate the closure property and the group structure. That is, if $x = x(f, g)$ and $y = y(f, g)$, then $\hat{x} = \hat{x}(\hat{f}, \hat{g})$, and $\hat{y} = \hat{y}(\hat{f}, \hat{g})$. One way to do this is to impose a condition whereby all the possible orbits have the same tangent vector at each point, i.e. $(\xi, \eta) = (0, 1)$, which leads to the general canonical system:

$$\xi(x, y)\frac{\partial}{\partial x} f(x, y) + \eta(x, y)\frac{\partial}{\partial y} f(x, y) = 0 \tag{7.13}$$

$$\xi(x, y)\frac{\partial}{\partial x} g(x, y) + \eta(x, y)\frac{\partial}{\partial y} g(x, y) = 1. \tag{7.14}$$

As canonical coordinates are not unique, Eqs. 7.13 and 7.14 are fulfilled by several functions f and g. Choosing simple f and g functions is convenient. Consequently,

$$r\frac{\partial}{\partial x} f(x, y) + \frac{\partial}{\partial y} f(x, y) = 0$$

$$r\frac{\partial}{\partial x} g(x, y) + \frac{\partial}{\partial y} g(x, y) = \frac{1}{\delta}, \tag{7.15}$$

Symmetry invertibility, i.e. Curve 1 $\rightleftarrows$ Curve 2, is another important constraint. Evidently, the tangent vectors in Eq. 7.14 cannot be zero simultaneously. Applying the chain rule on Eq. 7.7 gives the equivalent derivative operator, D_x, that still meets the closure property in the new system i.e.

$$\frac{d\hat{y}}{d\hat{x}} = \frac{D_x\hat{y}}{D_x\hat{x}} = \frac{D_x f}{D_x g}. \tag{7.16}$$

where $D_x = (\partial_x + y'\partial_y)$ in first-order, and $\mathcal{O}(\varepsilon^2)$. The solution of the ODE is obtained by solving the characteristic equation, $Q(x, y) = (\eta\text{-}\xi y') = 0$ i.e.

$$\frac{dy}{dx} = \frac{\eta(x, y)}{\xi(x, y)} = \frac{\delta}{\delta r} = \frac{1}{r}, \tag{7.17}$$

for $r \neq 0$. In the context of Noether's first theorem, Eq. 7.17 implies that r is a conserved quantity in this symmetry. That is, the characteristics exhibit the same resistance r in the vicinity of ε. Although r may vary with an external parameter e.g. illumination intensity, temperature, doping, etc., it varies identically near ε for all the characteristics in the same diode. Equation 7.17 has the general solution $y = (x/r+m)$, where m is an arbitrary constant.

The ODE in Eq. 7.17 is still physically meaningful since dy/dx denotes the reciprocal of resistance (conductance), y is the current, and x is the applied voltage bias. Also, there are no singular solutions since $r > 0$ for real devices. We can arbitrarily choose a canonical coordinate (f, g) in terms of m, e.g.

$$m = f = y - \frac{x}{r}, \tag{7.18}$$

This selection satisfies the first expression in Eq. 7.15 and f is invariant to the transformations. Then

$$g(f, x) = \int \frac{dx}{\xi(y, r)}\bigg|_r = \int \frac{dx}{\delta r} = \frac{x}{\delta r}. \tag{7.19}$$

Therefore, a possible canonical coordinate is

$$(f, g) = \left(y - \frac{x}{r}, \frac{x}{\delta r}\right), \quad \text{or} \quad (x, y) = \left(g\delta r, \ f + g\delta\right). \tag{7.20}$$

Equation 7.16 allows us to write the transformed derivative:

$$\frac{dg}{df} = \frac{g_x + \hat{y}' g_y}{f_x + \hat{y}' f_y} = \frac{(1/\delta r)}{(-1/r) + \hat{y}'}. \tag{7.21}$$

Substituting x and y from Eq. 7.20 into Eqs. 7.5 and 7.6 and simplifying the resulting expressions gives $\hat{\beta}$ and $\hat{y}'$ in terms of only one canonical coordinate i.e.

$$\hat{\beta} = \left[(b - c)e^{brf} + c\right]e^{-crf}, \tag{7.22}$$

$$\text{and} \quad \hat{y}' = \frac{\left[(b - c)e^{brf} + c\right]e^{-crf}}{p + r\left[(b - c)e^{brf} + c\right]e^{-crf}}. \tag{7.23}$$

Substituting Eq. 7.23 into Eq. 7.21 and simplifying then gives the transformed ODE in only one canonical variable:

$$\frac{dg}{df} = -\frac{1}{\delta} + \left(-\frac{r}{p\delta}\right)\left\{(b - c)e^{brf} + c\right\}e^{-crf}. \tag{7.24}$$

This result can be integrated easily w.r.t f to give the MS equation in the chosen canonical space:

$$g(f) = -\frac{f}{\delta} + \frac{1}{p\delta}\left[1 - e^{brf}\right]e^{-crf}. \tag{7.25}$$

The final check is whether the original Eq. 2.26 can be obtained from Eq. 7.25. We see that the direct substitution of f after multiplying through by δ and setting $g\delta = (y - f)$ from Eq. 7.20 then gives the original equation:

$$y = \frac{1}{p}\left[1 - e^{brf}\right]e^{-crf}, \tag{7.26}$$

remembering that $f=(y - x/r)$.

7.4.2 Case II—Bias-Dependent Barrier Height

Most literature [14, 15] take Φ as being linear with bias x i.e.

$$\Phi(x) = \Phi_0 + \alpha x, \tag{7.27}$$

where Φ_0 is the zero-bias barrier height and α is a constant bias-coefficient of barrier height (in eV/V). Differentiating Eq. 7.1 w.r.t x while noting that $\Phi' = \alpha$, gives:

$$y' = \frac{\left[(b + \alpha b - c)e^{-b(x-yr)} - (\alpha b - c)\right]e^{c(x-yr)}}{se^{b\Phi_0} + r\left[c + (b - c)e^{-b(x-yr)}\right]e^{c(x-yr)}} \tag{7.28}$$

$$\text{or} \quad y' = \frac{\theta(x, y)}{p + r\zeta(x, y)}, \tag{7.29}$$

where $p = s\exp(b\Phi_0)$. Equation 7.28 reduces to Eq. 7.4 when $\alpha = 0$, where $\theta = \zeta = \beta$. Furthermore, the symmetry condition for Eq. 7.28 still holds:

$$\hat{y}' = \frac{\hat{\theta}(\hat{x}, \hat{y})}{p + r\hat{\zeta}(\hat{x}, \hat{y})} \tag{7.30}$$

and is unique only for the transformations in Eq. 7.9. This suggests that the results preclude local variations around ε (or more accurately $\Delta\varepsilon$). Such is the case when, for example, the ideality factor n also depends on bias [18]. Since Δx evidently does not significantly affect n around $\Delta\varepsilon$, $n(x)$ can be estimated over the entire bias range by incrementally increasing x.

7.4.3   Uniqueness of the Symmetry

One can ask whether the above ODE has many, or just one unique, symmetry. Starting at Eq. 7.4

$$y' = \omega(x, y), \tag{7.31}$$

we can try arbitrary, smooth functions, or ansatzes, for the tangent vector field (ξ, η). Ansatzes can give insights into analytically challenging problems. Imposing the linearization in Eq. 7.10 on Eq. 7.4 yields the so-called *linearized symmetry condition*, or LSC [19–21]. We begin at Eq. 7.4 and get the LSC

$$\eta_x + (\eta_y - \xi_x)\omega - \xi_y\omega^2 = \xi\omega_x + \eta\omega_y, \tag{7.32}$$

in terms of the partial derivatives w.r.t the subscripts. This condition can speed up finding symmetries, thereby easing solving the ODE.

Theorem 7.2 *The I–V symmetry arising from Eq. 7.9 is unique.*

Proof It suffices to show that all non-trivial ansatzes $(\xi(x, y), \eta(x, y))$ give Eq. 7.9. A frequently used ansatz [20] is

$$\xi = \alpha(x), \quad \eta = \Gamma(x)y + \gamma(x), \tag{7.33}$$

where α, Γ, and γ are unknown but smooth and arbitrary functions. Then, using Eq. 7.33 into Eq. 7.32 with $\omega(x, y)$ defined through Eqs. 7.4 and 7.5 leads to

$$\gamma'(x) = \alpha'(x) = 0, \quad \Gamma = 0, \quad \alpha(x) = r\gamma(x). \tag{7.34}$$

Both α and Γ are thus found to be constants i.e. $(\xi, \eta) = (r\gamma, \gamma)$. This is the same result in Eq. 7.9. Another well-known ansatz that is used with rotations, translations, scalings, and so on [19, 20] is

$$\xi = c_1 x + c_2 y + c_3, \quad \eta = c_4 x + c_5 y + c_6 \tag{7.35}$$

where $\{c_1, \ldots, c_6\}$ are arbitrary constants. Again, one finds $(\xi, \eta) = (rc_6, c_6)$. Therefore, Eq. 7.9 arises regardless of the choice of ξ and η.

7.5   Series Resistance

Figure 7.4 illustrates the translation symmetry action on the I–V characteristics of the MS diode over the $\pm x$ bias regions. This wider range is a clear improvement over other methods.

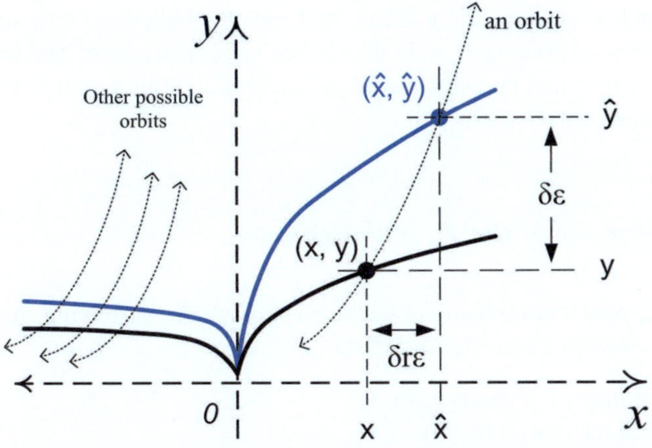

Fig. 7.4 Thematic representation of the symmetry action of Eq. 7.7 on the points (x, y) and $(\hat{x}, \hat{y})$ that are located on the I–V characteristics of the MS diode. The action is illustrated in the $+x$, forward bias region, but is equally valid in the $-x$, or reverse bias region

7.5.1 Newtonian Mechanics Analogy

We now return to the Newtonian mechanics analogy of the infinitesimal ε. As $\varepsilon \to 0$, we see that ξ and η respectively represent the instantaneous rates of change of applied bias and current wr.t. ε. For real I–V characteristics, we can talk of averaged values around ε. That is, setting $\delta\varepsilon \to \Delta\varepsilon$ in Eq. 7.9 implies that

$$\hat{x} = x + r\Delta\varepsilon, \quad \text{and} \quad \hat{y} = y + \Delta\varepsilon. \tag{7.36}$$

The graphical determination of r then becomes trivial on an experimental data set:

$$r = \frac{\hat{x} - x}{\hat{y} - y} = \frac{r\Delta\varepsilon}{\Delta\varepsilon}. \tag{7.37}$$

The compensation for resistance on an I–V data set requires that we first locate the points (x, y) and $(\hat{x}, \hat{y})$. The separation $(x - \hat{x})$ denotes the voltage step of the measurements. In the experimental validation below, the instrumentation acquires data in voltage bias steps of 0.1 V. In the ensuing analysis, we can arbitrarily choose a small enough $(x - \hat{x})$ e.g. 0.2 V, so that Eq. 7.37 effectively spans 3-data points while still meeting the condition of Eq. 7.36 i.e. $\Delta\varepsilon \to 0$. The modified bias at x can then be calculated from

$$(x - yr)|_x \equiv (V - IR_s)|_V, \quad \text{where } I = \max.\{y, \hat{y}\}|_\varepsilon.$$

That is, at each x we generate a datum that satisfies both Eqs. 7.18 and 7.26, thus giving a new set of points (f, y) in the chosen canonical space. The bigger current in the vicinity of ε can be used to find the maximum compensation to the applied bias at the point.

7.5.2 Summary of the R_s Algorithm

Despite the apparent complexity of the new approach, the following steps outline its application to compensate empirical data.

- locate the points (x, y) and $(\hat{x}, \hat{y})$
- calculate $r = (\hat{x} - x)/(\hat{y} - y)$
- calculate the modified bias at x using r:

$$(x - yr)|_x \equiv (V - I R_s)|_V, \quad \text{where } I = \max.\{y, \hat{y}\}|_\varepsilon.$$

- adjust I for each voltage point centered on the range $(\hat{x}\text{-}x)$.

Figure 7.5 shows how the instantaneous values of R_s at a bias $V = x_n$ and current y_n can be calculated using adjacent experimental data points in all regions of bias.

For real devices, serialized curves have temperature-dependent symmetry. However, once the symmetry is established, a simple and direct method based on a near-identity plane transformation can be applied over the entire applied bias region.

Many semilog I–V methods approximate the MS equation by discarding exponential terms. This narrows the bias range of application and generally gives less reliable parameter estimates. In the absence of a rigorous error analysis, the true impact of such approximations is difficult to ascertain.

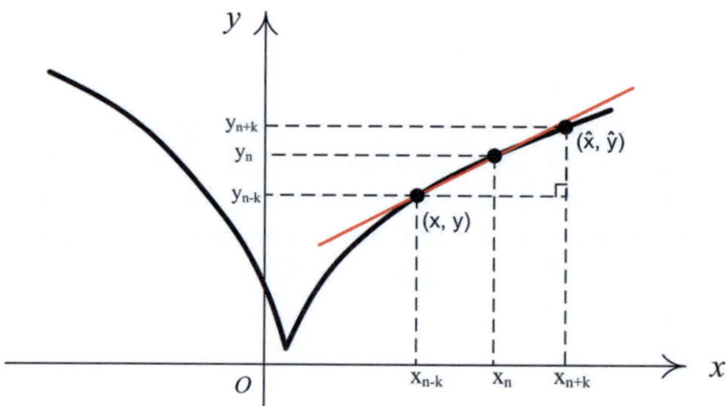

Fig. 7.5 An illustration showing how instantaneous R_s values can be calculated using data from an experiment

The proposed approach avoids these difficulties. The following cases were tested using the method:

(a) (x, y) and $(\hat{x}, \hat{y})$ are chosen on two different characteristics
(b) $(\hat{x}, \hat{y})$ is located on the same characteristic as (x, y) but with $\hat{x} > x$.

Since the canonical is known, the logarithm plot of Eq. 7.26 using the data set can be compared directly with the results of R_s, n, and Φ_0 from the above two test cases. The results are compared with the Cheung-Cheung method [22].

7.5.3   Method Results

The I–V plots in Fig. 7.6 are based on a surveyed MSM diode over 0–100 mW/cm^2 illuminations [32, 33]. It shows both raw uncompensated and R_s-compensated plots.
   Table 7.1 shows the results of the method from the two test cases.
   Figure 7.7 shows that in both regions of bias:

$$R_s = R_0 e^{-\alpha V_a},\tag{7.38}$$

where V_a is applied bias, and α is a constant. The calculated R_0 are 61.4 and 13.2 kΩ respectively. Many prior works using CV measurements [23] support a decreasing R_s with bias. The new method hints at a dependence of α on illumination and bias region i.e. it increases linearly in the first region of bias, and decreases linearly in the second. Similar trends have been reported [1]. Furthermore, plots like Fig. 7.7 often

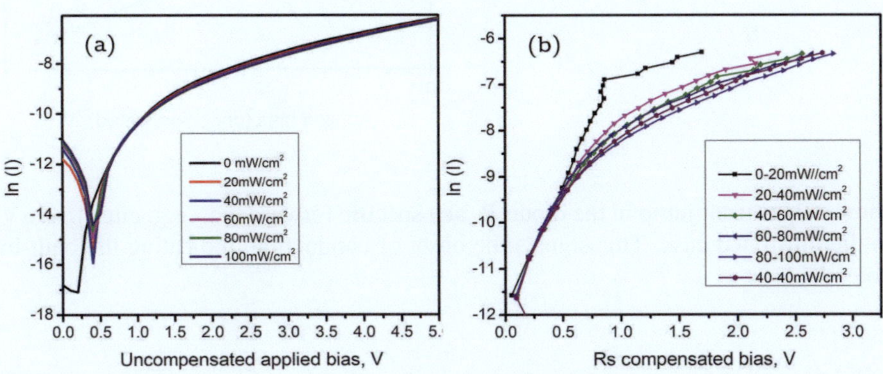

Fig. 7.6   Illustration of the impact of the new R_s compensated method using real experimental data on a published diode. In **a** uncompensated plots, and **b** after R_s compensation. The 40–40 mW/cm^2 plot is the case where Curve 1 = Curve 2. Evidently, the new method leads to a clear separation of the trends in the different characteristics that are suppressed due to the presence of the series resistance

Table 7.1 Calculated n and Φ_b using the proposed method in the two identified regions. The percent variances in Φ_b w.r.t average Φ_b in the first and second bias regions are 0.007 and 0.044, respectively. The constants are $A^* = 32$ A/K^{2}cm^2, $A = 1$ mm, and $T = 300$ K

P (mW/cm^2)	Bias (0.1–1.0 V)			Bias (1.0–4.8 V)		
	n	α^3 (/V)	Φ_b	n	α^4 (/V)	Φ_b
0–20	7.1	−2.027	0.567	47.7	−0.417	0.458
0–100	7.3	−^{1}	0.566	–	–	–
20–40	8.9	−1.914	0.561	37.9	−0.511	0.483
20–100	5.9	–	0.575	43.9	–	0.468
40–60	10.3	−1.832	0.557	39.9	−0.530	0.486
60–80	8.9	−1.679	0.560	33.9	−0.578	0.495
80–100	9.7	−1.806	0.559	37.6	−0.591	0.497
40–40^2	10.8	−1.914	0.557	39.2	−0.511	0.491

^{1}Not available
^{2}Curve 1 = Curve 2 test case
$^3\alpha = 3.51 \times 10^{-3} P - 2.03$
$^4\alpha = -2.22 \times 10^{-3} P - 0.41$

Fig. 7.7 Plots showing the test cases where Curve 1 $\neq$ Curve 2 (20–40, and 80–100 mW/cm^2), and Curve 1 = Curve 2 (40–40 mW/cm^2). In all cases, the resistance at a given bias depends to a small extent on the illumination

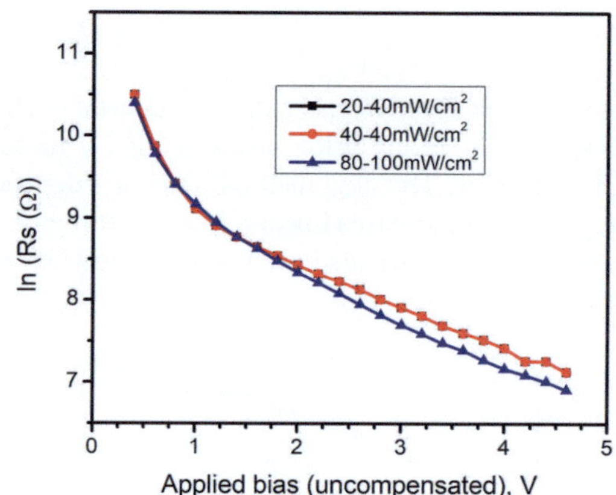

show a significant jump in the diode R_s at a specific forward bias, e.g. circa 0.476 V in the illustrated case. This signals the onset of conduction, indicating the built-in potential.

7.5.4   Reduced TE Equation

The effect of R_s compensation is tantamount to simplifying the TE equation. The barrier height can be calculated using Eq. 7.26. Denoting the R_s bias by z, with $w = (1/p)$, Eq. 7.26 can be written:

Table 7.2 Calculated R_s, n and Φ_b using the Cheung-Cheung method in the two identified regions. The variances in Φ_b in percent of average Φ_b in Regions A and B are 0.088 and 0.004, respectively. The bias regions are A: 0.6–1.2 V, and B: 2.3–4.8 V. R_s are in $k\Omega$, and Φ in (eV)

	$dV/d\ln(I)$ vs I				$H(I)$ vs I				Averages	
	A		B		A		B			
	n	R_s	n	R_s	R_s	Φ_b	R_s	Φ_b	$R_s{}^1$	$R_s{}^2$
0	6.4	4.07	32.2	0.64	3.26	0.785	0.72	0.524	3.66	0.68
20	4.3	6.11	34.5	0.57	6.37	0.712	0.63	0.532	6.24	0.60
40	4.1	6.84	34.6	0.56	6.79	0.723	0.62	0.534	6.82	0.59
60	3.4	7.32	35.6	0.52	7.67	0.753	0.58	0.533	7.49	0.56
80	3.6	7.67	35.7	0.50	7.95	0.740	0.56	0.535	7.81	0.53
100	3.4	7.46	34.3	0.56	7.86	0.748	0.61	0.537	7.66	0.58

[1] A, $dV/d\ln(I)$ function
[2] B, $H(I)$ function

$$\ln y = cz + \ln w. \tag{7.39}$$

That is, a plot of $\ln y$ versus z in both bias regions is a straight line. The gradient c leads to n, since $n = b/c$, where $b = q/kT$. The intercept leads to

$$\text{intercept} = AA^*T^2 e^{-b\Phi_b}. \tag{7.40}$$

Furthermore, it is apparent that the results of the test case (b) i.e. Curve 1 = Curve 2 when $P = 40\,\text{mW/cm}^2$, are practically identical to test case (a) i.e. Curve 1 $\neq$ Curve 2. Test case (b) is thus sufficient for good accuracy.

Table 7.2 summarises the results of the Cheung-Cheung method on the surveyed Schottky photodiode. The method also suggests that the two bias regions influence the series resistance, being generally lower for higher bias. In Table 7.2 both regions show an increase in resistance with illumination. From a physical perspective one expects that the conductance should increase with illumination i.e. the resistance decreases. Figure 7.8 shows the semilog plots for the two regions of bias. The calculated n and Φ_b using the two methods are generally comparable (Fig. 7.9).

Figure 7.10 is a comparison plot showing the typical values of R_s determined using the reduced TE method and the Cheung-Cheung method. The plots represent the same ambient conditions of illumination and temperature.

The figure shows only the general trends in the Cheung-Cheung method results. It is accepted that each of its functions is applicable in at least two separate narrow bias regions, and its calculated parameters also vary widely on these ranges. In actual plots, it is evident that the reduced TE calculates R_s over a wider range which incidentally covers the Cheung-Cheung method, such as was recently shown by Ocaya et al. [24].

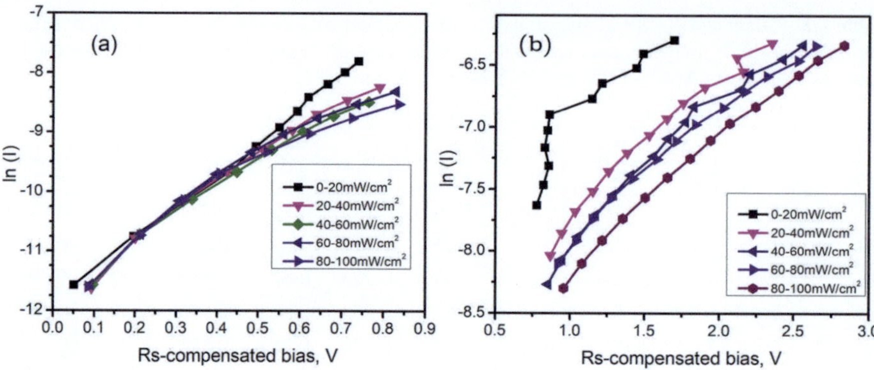

Fig. 7.8 Comparison of the semilog plots in the low and high compensated bias regions. In (**a**), the plot is generated in part of the reverse bias, as is apparent in Fig. 7.6

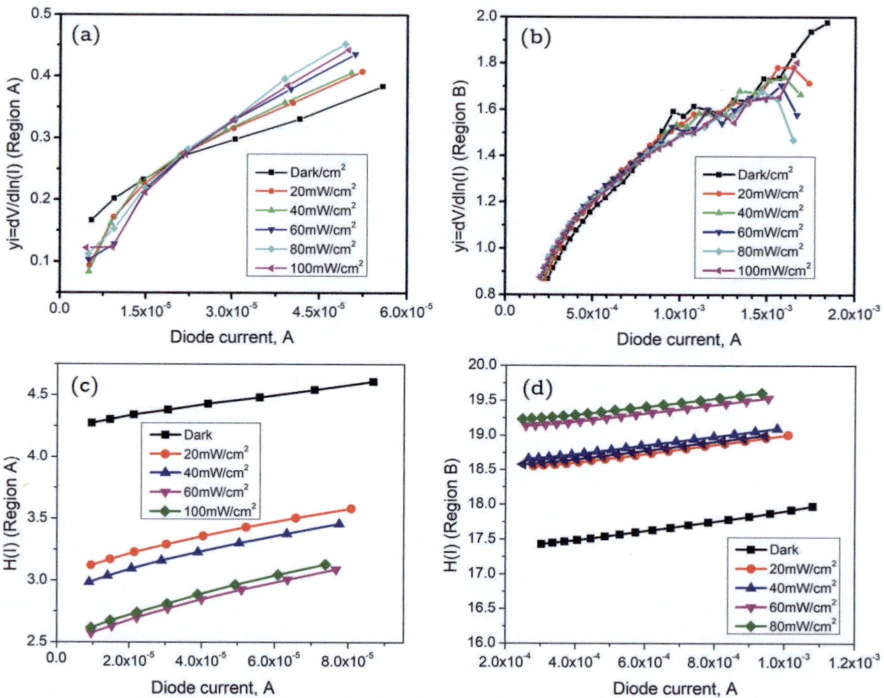

Fig. 7.9 Cheung-Cheung function plots. The bias range is 0.6–1.40 V in (**a**), (**c**), and 2.3–3.9 V in (**b**) and (**d**). The functions showed low linearity outside these ranges

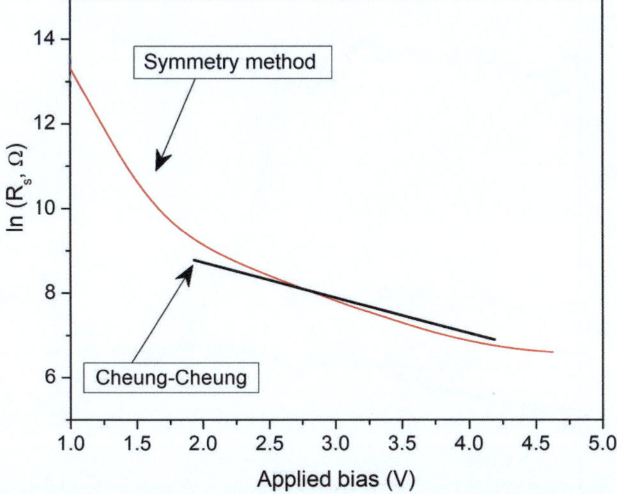

Fig. 7.10 A plot showing a side-by-side comparison of the symmetry-based reduced TE method with the Cheung-Cheung method. The plots based on the former method are a small, approximated subset of the reduced TE method. This suggests that the symmetries approach is the superior method

7.5.5 Comparison with AC Impedance Measurements

Existing I–V methods can give R_s values that differ by several orders of magnitude from AC impedance methods, a noted discrepancy that is not adequately justified in the literature.

Figure 7.11 shows the calculated resistance variation-bias variation for a surveyed diode [25] using the suggested method plotted with the AC measurements at 10 kHz. Under dark and 2V bias, the Cheung-Cheung method reported 167 Ω, against 11 kΩ at 10 kHz in the C-V method, and 14.5 kΩ in the new method. The latter two measurements are two orders of magnitude higher. However, the C-V measurements exhibit the trend of decreasing series resistance with measurement frequency. There-fore, in measurements at frequencies lower than 10 kHz i.e. approaching DC, R_s is marginally higher than 11 kΩ. The plot suggests that R_s decreases with increasing reverse bias, such that the reverse current should increase with reverse bias. This is not observed in the I–V plots. Therefore, the new method better agrees with the C-V method than the Cheung-Cheung method but is unparalleled in supporting the intuitive behavior seen in real I–V plots.

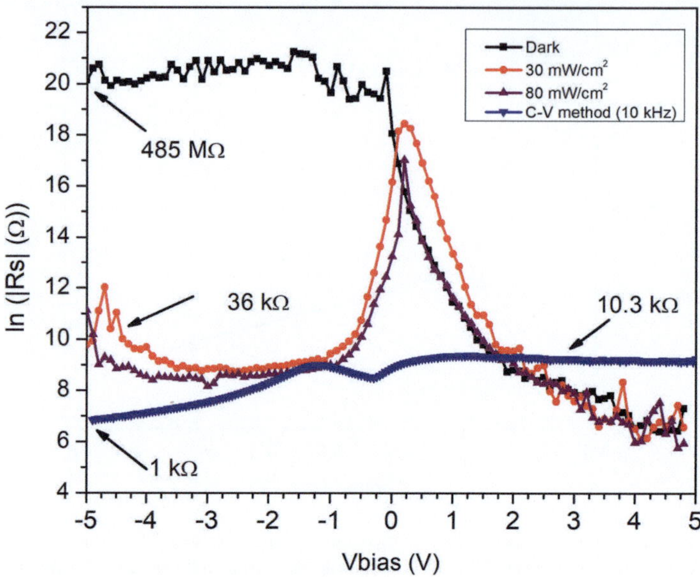

Fig. 7.11 Plot of the resistance variation with applied bias under dark, 30, and 80 mW/cm² illumination intensities using the new method and the AC impedance method. The results of the Cheung-Cheung method cannot be plotted on the same graph

7.5.6   Method Impact on Other Diode Parameters

Next, we report on the impact of this new method on other important diode parameters. We limit our present scope to the diode photoconductive response and the interface state density.

7.5.6.1   Photoconductivity Response

A photoconductive material with an exponential distribution of carrier trap states and a bandgap that is less than the incident photon energies has an electron density given by:

$$N \propto e^{\frac{E_c - E_t}{E_c}}, \tag{7.41}$$

where $E_c = kT_c$ is a temperature-dependent constant for the material, and the thermal energy is $E_t = kT$. Rose [26, 27] derived the free electron density in terms of the illumination-dependent photogeneration rate G, as

$$n \propto G^\gamma, \quad \text{where} \quad \gamma = \gamma(T) = \frac{T_c}{T + T_c}. \tag{7.42}$$

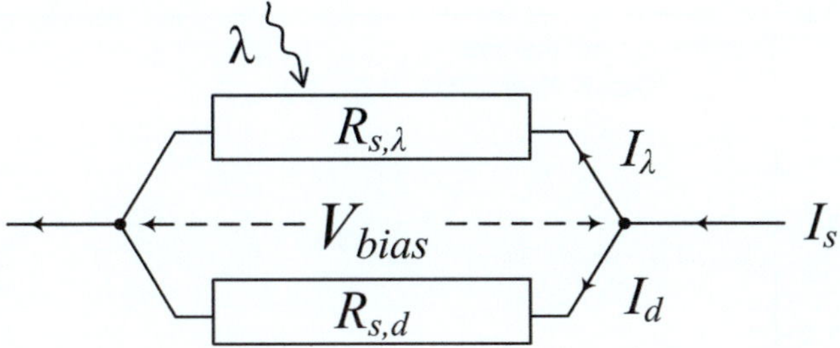

Fig. 7.12 The total diode current, I_d, consists of I_d, and the photocurrent, I_λ. The dark and illumination R_s are $R_{s,d}$ and R_λ, respectively

The exponent γ indicates the dominant carrier recombination mechanism in the material. Equation 7.41 suggests that the recombination current increases with the population of trap states, N. This requires that $(E_c > E)$ i.e. $(T_c > T)$. The steady-state photoconductive current is the net recombination current. For many photoconductive materials where the distribution of trap states is exponential, $0.5\gamma < 1$. This corresponds to $0 < T < T_c$. The reverse bias current in a photodiode may be modeled by the equivalent circuit in Fig. 7.12.

Since $R_{s,d} >> R_\lambda$, as in Fig. 7.11, it follows that $R_s \sim R_\lambda$.

At the illumination P the photo-conductivity of such materials has been shown to have form [26, 34]

$$\sigma_\lambda = \sigma_0 P^\gamma, \tag{7.43}$$

where σ_0 is a material-dependent constant. If we relate Eq. 7.43 to the photo-resistivity, ρ_λ, then, in terms of a constant ε that ordinarily has dimension 1/L we can write $R_\lambda = \rho_\lambda \varepsilon$. Then, since $\sigma_\lambda = 1/\rho_\lambda$, Eq. 7.43 could be written as:

$$\frac{\varepsilon}{R_\lambda} = \frac{\alpha}{\alpha_v} P^\gamma, \quad \text{or} \quad \frac{(\varepsilon\alpha_v)}{R_\lambda} = \frac{V_{bias}}{R_s} = \alpha P^\gamma, \tag{7.44}$$

in terms of a bias-dependent term, α_v, such that $\alpha = \alpha_v \sigma_0$. In short, the present method indicates that the photo-conductivity is also bias-dependent. There is ample support for this in the literature [28, 29].

Therefore, for a diode in which the trap states distribution is exponential ($0.5 < \gamma < 1$), one expects that at a given bias, the plot of $1/R_s$ versus P^γ is a straight line through the origin, with gradient α. The constant γ can be estimated at a given bias from

$$\gamma = \frac{\log\ (R_{s,1}/R_{s,2})}{\log\ (P_2/P_1)}, \tag{7.45}$$

where the corresponding series resistances at the two illuminations P_1 and P_2 are $R_{s,1}$ and $R_{s,2}$, respectively. The constant α can be found using Eq. 7.43.

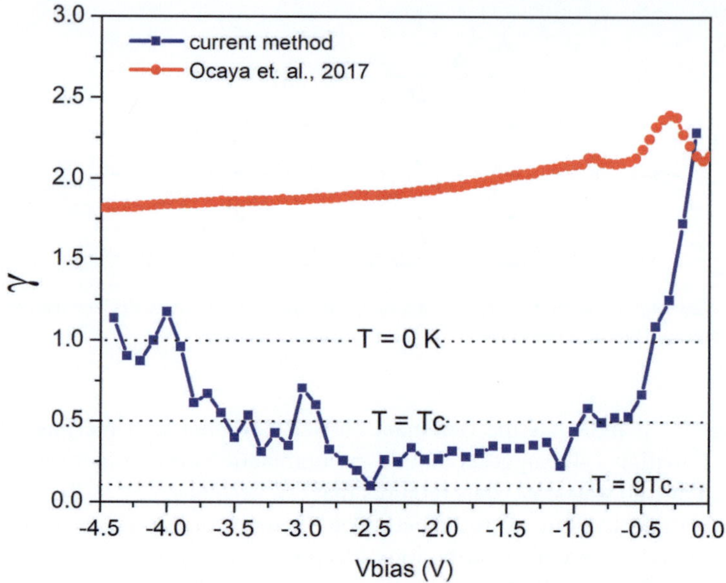

Fig. 7.13 Calculated power exponent as a function of applied bias in the photoconductive mode of a surveyed MSM photodiode. Ordinarily, the diode is used in the temperature region between 0 K and T_c

Figure 7.13 shows the plot of γ versus applied bias for a surveyed diode [25]. The plot suggests that the photo-conductivity follows the power-law, but the distribution of trap states is exponential over a narrow range of applied bias. Other photo-conducting materials are known to exhibit values of γ outside the $0.5 < \gamma < 1$ range under specific conditions [30, 31]. The disagreements in the plots in Fig. 7.13 show suggest that the earlier model does not adequately describe the photo-response of the diode.

7.5.6.2   Interface state density

Figure 7.14 shows the plot of R_s with an applied bias for a surveyed diode using the presented method.

The C-V is another popular method for barrier height measurement through the AC conductance, G, and depletion layer capacitance, C_d. It arrives at other parameters, such as the built-in potential, V_{bi}, and the interface state density, N_i. However, measured R_s values are strongly bias sensitive and frequency reactive [17]. This increases the method complexity and uncertainty [35, 36].

The C-V method results of N_i are shown in Table 7.3 for a surveyed diode.

It can be seen that the built-in potential from the C-V method is almost six times larger compared to the new method. The bandgap from the proposed method agrees

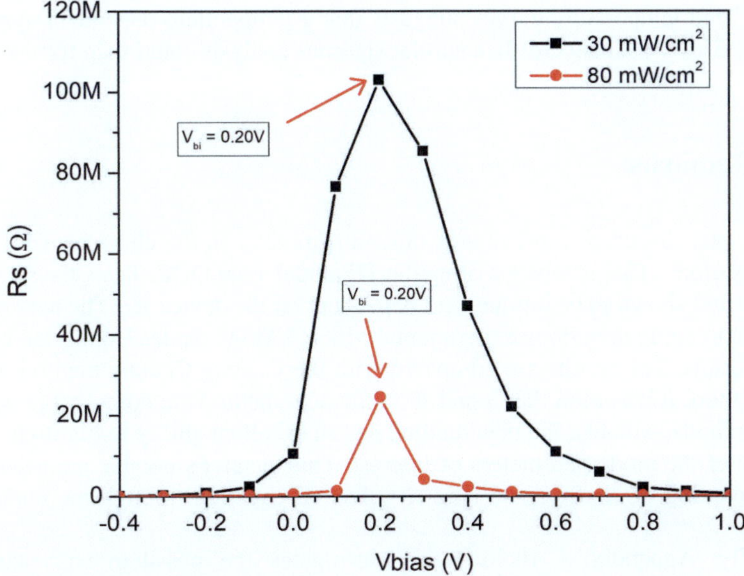

Fig. 7.14 Calculated variation of series resistance with applied bias at two illumination levels for an experimental diode [25]. The built-in potential is 0.2 V from either plot

Table 7.3 Calculated non-ionized acceptor concentrations using the C-V method and the new method [17, 25]

Method	N_i (/cm³)	V_{bi} (V)	Φ (eV)
C-V	1.71×10^{14}	1.287	1.002
Modified TE	8.68×10^{11}	0.200	0.641

well with the Cheung-Cheung method but their R_s values do not [16, 25, 32, 37]. It is known that the non-ionized acceptor concentration in p-Si at room temperature is of the order of $\sim 10^{10}$ /cm³. Thus, the method produces a value that is closer to what is expected than the C-V method.

7.5.7 *Other Potential Symmetries*

In passing, there may exist other unique symmetries w.r.t to different external parameters. In the specific case of temperature being that variable, for instance, then an ODE of the form $z' = \omega(x, y, z)$ could be attempted in terms of $x (= V)$, $y (= T)$, and $z (= I)$. One might designate a function $\ln(z)$ that is transformed along y under constant x, or along x as y remains constant. The ODE linearization would then involve tangent vector fields for the three variables. Inspection of the proliferation of serialized temperature curves suggests that a temperature-dependent symmetry should exist [33]. Thenceforth, a similar rigorous analysis could then follow.

7.6 Remarks

This chapter discussed a previously unseen symmetry in the characteristics of the Schottky diode. This involves writing the TE model as an ODE. Then, the symmetry is found and shown to be unique, and dependent on the device R_s. The new method is shown to apply to various experimental MS and MSM diodes, supported by published results. The results are compared with the Cheung-Cheung method and the C-V method. It was seen that n and Φ in the new method are comparable with the other methods. Notably, the new method has an excellent ability to highlight trends in many of the diode parameters of interest. This chapter's notable contribution is the decoupling of R_s from the hard-to-solve TE equation through the underlying symmetry.

Finally, Appendix A rhetorically interrogates the possible existence of a temperature-based symmetry. Clearly, there is much scope for this line of research into more efficient methods that could exploit an underlying symmetry, if it exists.

References

1. H.E. Lapa, A. Kökce, D. Aldemir, A.F. Özdemir, Ş. Altindal, Effect of illumination on electrical parameters of Au/(P3DMTFT)/n-GaAs Schottky barrier diodes. Indian J. Phys. 1–8 (2019). https://doi.org/10.1007/s12648-019-01644-y
2. G. Güler, Ö. Güllü, Ş Karataş, Ö.F. Bakkaloğlu, Analysis of the series resistance and interface state densities in metal semiconductor structures. J. Phys.: IOP Conf. Ser.**153**, 012054 (2009)
3. F. Hernandez-Ramirez, A. Tarancon, O. Casals, E. Pellicer, J. Rodriguez, A. Romano-Rodriguez, J.R. Morante, S. Barth, S. Mathur, Electrical properties of individual tin oxide nanowires contacted to platinum electrodes. Phys. Rev. B **76**(8), 085429 (2007)
4. Z. Wang, W. Zang, Y. Shi, X. Zhu, G. Rao, Y. Wang, J. Chu, C. Gong, X. Gao, H. Sun, others, Extraction and analysis of the characteristic parameters in back-to-back connected asymmetric schottky diode, Physica Status Solidi (a) **217**(8), 1901018 (2020)
5. R. Nouchi, Extraction of the Schottky parameters in metal-semiconductor-metal diodes from a single current-voltage measurement. J. Appl. Phys. **116**(18) (2014). https://doi.org/10.1063/1.4901467.
6. İ. Taşçıoğlu, S.O. Tan, F. Yakuphanoğlu, Ş. Altındal, Effectuality of barrier height inhomogeneity on the current-voltage-temperature characteristics of metal semiconductor structures with CdZnO interlayer. J. Electron. Mater. **47**(10), 6059–6066 (2018)
7. E. Noether, Invariant variation problems. Transp. Theory Stat. Phys. **1**(3), 186–207 (1971). https://doi.org/10.1080/00411457108231446
8. J.H. Werner, Schottky barrier and pn-junction IV plots—small signal evaluation. Appl. Phys. A **47**(3), 291–300 (1988)
9. E.H. Rhoderick, R.H. Williams, *Metal-Semiconductor Contacts* (Clarendon Press, 1988)
10. S. Chand, J. Kumar, Electron transport and barrier inhomogeneities in palladium silicide Schottky diodes. Appl. Phys. A **65**(4), 497–503 (1997)

11. D.E. Yildiz, Ş Altindal, H. Kanbur, Gaussian distribution of inhomogeneous barrier height in Al/SiO2/p-Si Schottky diodes. J. Appl. Phys. **103**(12), 124502 (2008). https://doi.org/10.1063/1.2936963

12. A.M. Chowdhury, R. Pant, B. Roul, D.K. Singh, K.K. Nanda, S.B. Krupanidhi, Double Gaussian distribution of barrier heights and self-powered infrared photoresponse of InN/AlN/Si (111) heterostructure. J. Appl. Phys. **126**(2), 025301 (2019)

13. S. Khanna, S. Neeleshwar, A. Noor, Current-voltage-temperature (IVT) characteristics of Cr/4H-SiC Schottky diodes. J. Electron Dev. **9**, 382–389 (2011)

14. H. Durmuş, Ü. Atav, Extraction of voltage-dependent series resistance from IV characteristics of Schottky diodes. Appl. Phys. Lett. **99**(9), 093505 (2011)

15. V. Mikhelashvili, G. Eisenstein, R. Uzdin, Extraction of Schottky diode parameters with a bias dependent barrier height. Solid-State Electron. **45**(1), 143–148 (2001)

16. H. Zhang, X. Zhang, C. Liu, S.-T. Lee, J. Jie, High-responsivity, high-detectivity, ultrafast topological insulator Bi_2Se_3 /silicon heterostructure broadband photodetectors. ACS Nano **10**(5), 5113–5122 (2016). https://doi.org/10.1021/acsnano.6b00272PMID: 27116332

17. R.O. Ocaya, A.G. Al-Sehemi, A. Al-Ghamdi, F. El-Tantawy, F. Yakuphanoğlu, Organic semiconductor photosensors. J. Alloys Compd. **702**, 520–530 (2017). https://doi.org/10.1016/j.jallcom.2016.12.381

18. V. Mikhelashvili, G. Eisenstein, The influence of image forces on the extraction of physical parameters in Schottky barrier diodes. J. Appl. Phys. **86**(12), 6965–6969 (1999)

19. R.O. Ocaya, *Introduction to Control Systems Analysis Using Point Symmetries: An Application of Lie Symmetries* (AuthorHouse, UK, 2016)

20. P.E. Hydon, *Symmetry Methods for Differential Equations: A Beginner's Guide* (Cambridge University Press, 2000)

21. N.K. Ibragimov, Group analysis of ordinary differential equations and the invariance principle in mathematical physics (for the 150th anniversary of Sophus Lie). Russian Math. Surv. **47**(4), 89–156 (1992)

22. S.K. Cheung, N.W. Cheung, Extraction of Schottky diode parameters from forward current-voltage characteristics. Appl. Phys. Lett. **49**(2), 85–87 (1986)

23. A. Mekki, R.O. Ocaya, A. Dere, A.A. Al-Ghamdi, K. Harrabi, F. Yakuphanoğlu, New photodiodes based on graphene-organic semiconductor hybrid materials. Synth. Metals **213**, 47–56 (2016)

24. R.O. Ocaya, I. Erol, A.G. Al-Sehemi, A. Dere, A.A. Al-Ghamdi, F. Yakuphanoğlu, ZnO-doped PFPAMA: a novel transparent conducting polymer for fast photodiodes. J. Mater. Sci.: Mater. Electron. **33**(32), 24803–24818 (2022). https://doi.org/10.1007/s10854-022-09192-8

25. R.O. Ocaya, A. Al-Ghamdi, K. Mensah-Darkwa, R.K. Gupta, W. Farooq, F. Yakuphanoglu, Organic photodetector with coumarin-adjustable photocurrent. Synth. Metals **213**, 65–72 (2016). https://doi.org/10.1016/j.synthmet.2016.01.002

26. A. Rose, *Concepts in Photoconductivity and Allied Problems* (Interscience Publishers, 1963)

27. K.C. Kao (ed.), Chapter 7—Electrical Conduction and Photoconduction, in *Dielectric Phenomena in Solids* (Academic, San Diego), pp. 381–514 (2004). isbn: 978-0-12-396561-5. https://doi.org/10.1016/B978-012396561-5/50017-7

28. K. Konno, O. Matsushima, D. Navarro, M. Miura-Mattausch, Limit of validity of the drift-diffusion approximation for simulation of photodiode characteristics. Appl. Phys. Lett. **84**(8), 1398–1400 (2004)

29. G. Cao, H. Gong, H. Qiu, L. Kong, S.H. Hu, N. Dai, Bias-dependent photocurrent of $Hg_{1-x}Cd_x$ Te photodiodes. J. Appl. Phys. **98**(6), 064504 (2005)

30. R.H. Bube, others, *Photoconductivity of Solids* (RE Krieger Pub. Co., 1978)

31. F. Stöckmann, Photodetectors, their performance and their limitations. Appl. Phys. **7**(1), 1–5 (1975)

32. R.O. Ocaya, A.G. Al-Sehemi, A. Dere, A.A. Al-Ghamdi, F. Yakuphanoğlu, Electrical, photoconductive, and photovoltaic characteristics of a Bi_2Se_3 3D topological insulator based metal-insulator-semiconductor diode. Sens. Actuators A: Phys. **341**, 113575 (2022). https://doi.org/10.1016/j.sna.2022.113575

33. A. Hussain, Temperature dependent current-voltage and photovoltaic properties of chemically prepared (p) Si/(n) Bi2S3 heterojunction. Egypt. J. Basic Appl. Sci. **3**(3), 314–321 (2016)
34. S. Kazim, V. Ali, M. Zulfequar, M.M. Haq, M. Husain, Electrical transport properties of poly [2-methoxy-5-(2'-ethyl hexyloxy)-1, 4-phenylene vinylene] thin films doped with acridine orange dye. Physica B: Condens. Matter **393**(1–2), 310–315 (2007)
35. B. Gunduz, A.A. Al-Ghamdi, A.A. Hendi, Z.H. Gafer, S. El-Gazzar, F. El-Tantawy, F. Yakuphanoğlu, New Schottky diode based entirely on nickel aluminate spinel/p-silicon using the sol-gel spin coating approach. Superlattices Microstruct. **64**, 167–177 (2013). https://doi.org/10.1016/j.spmi.2013.09.022
36. A.N. Donald, *Semi-conductor Physics and Devices* (Tata McGraw Hill Education Private Limited, 2006)
37. W.C. Huang, T. Lin, C. Horng, C. Chen, Barrier heights engineering of Al/p-Si Schottky contact by a thin organic interlayer. Microelectron. Eng. **107**, 200–204 (2013)

Chapter 8
Artifical Intelligence Parameter Extraction Methods

Abstract In this chapter, we introduce a pioneering approach employing three distinct machine learning (ML) models: Artificial Neural Network (ANN), Logistic Regression (LR), and Decision Tree (DT). These models are employed for the comprehensive analysis and extraction of internal parameters of Schottky photodiodes (SPDs), all while operating in the absence of any prior knowledge of the intricate and nonlinear thermionic emission (TE) expression governing the device current. Our methodology is validated through extensive training, evaluation, and demonstration of the ML models across a dozen proprietary datasets. These datasets represent the current responses of graphene oxide (GO) doped p-Si Schottky barrier diodes (SBDs) under varying conditions, encompassing both different levels of GO doping (0, 1, and 10%) and illumination levels spanning from complete darkness (0 mW/cm^2) to 30 mW/cm^2. Notably, the predictions made by these models are consistent, even at an intensity of 60 mW/cm^2. The training dataset incorporates independently calculated values of barrier height (ϕ), ideality factor (n), and series resistance (R_s), all derived using the Cheung-Cheung method, for each diode. Subsequently, the models accurately predict these parameters at unspecified intensities, including 80 and 100 mW/cm^2 on the model development data and 5% and 20% GO doping levels, which were not part of the development dataset. Remarkably, the ANN achieves an outstanding training and test accuracy of 100% and 99.99%, respectively, across most datasets, while the LR and DT models attain a validation and test accuracy of 100%. This demonstrates the ML models' remarkable capacity to efficiently grasp the photodiodes' photo responses and predict the internal SBD parameters with exceptional accuracy, all without relying on any inherent comprehension of the thermionic emission (TE) equation for SBDs. Our proposed ML models hold the potential to significantly streamline the analysis phase in device development cycles and can be readily extended for application to diverse datasets across various domains.

Artificial Intelligence (AI) is the computer science field that aims to simulate human intelligence using agents that perform tasks like or exceeding humans. These tasks include learning, reasoning, problem-solving, perception, and language understanding. This requires that AI systems analyze vast data, adapt to different situations, make decisions, and improve their performance over time in ways similar to the cognitive functions of the human brain. AI systems are complemented by advances in computing power and techniques and are evolving at an unprecedented rate [8], with

R. Ocaya, *Extraction of Semiconductor Diode Parameters*,
https://doi.org/10.1007/978-3-031-48847-4_8

applications now covering many areas, including the physical and applied sciences [9]. The term "AI" broadly encompasses a number of approaches and subfields, particularly:

1. Machine Learning (ML)—algorithms and statistical models that learn, predict, or make data-based decisions without being explicitly programmed with laws underlying the data. ML algorithms mimic human learning of specific data and establish meaningful connections between input and output variables. ML has the following subfields:

   - Deep Learning (DL)—neural networks with many interconnected layers termed deep neural networks (DNN). DL is particularly useful for image and speech recognition.
   - Reinforcement Learning (RL)—where the agent learns to make decisions while adjusting its behavior using reward or penalty-based feedback. RL may rely on human feedback.

2. Natural Language Processing (NLP)—agents that enable understanding, interpreting, and generating natural human language.
3. Computer Vision—agents enable the interpretation and understanding of visually presented world information i.e. images and videos.
4. Robotics—agents integrate AI with mechatronics in order to autonomously or semi-autonomously perform tasks in industrial settings, vehicles, droids, and drones.
5. Expert Systems—the agents use knowledge-based rules and reasoning engines to mimic the decision-making abilities of human experts.
6. Cognitive Computing—the agents attempt to emulate human cognitive functions primarily to complement humans in complex tasks.

ML has become a powerful impetus for scientific and engineering advancements despite being in its infancy. At present, however, it faces the handicap of relearning each new task from zero in contrast to the human brain's ability to transfer knowledge across different tasks without relearning [3]. This chapter cannot hope to be an exhaustive reference on AI techniques but has a narrow focus on illustrating, through examples, the newly discovered potential of ML to the challenging problem of parameter extraction from the TE equation.

8.1   Recent Applications of ML in Materials Research

Within the last five years, ML-based methods have been used increasingly to solve a variety of complex problems in materials science. For instance, Buratti et al. [1] suggested an ML approach called random forest regression [11] to predict performance and reliability limiting defect parameters and their energy levels in the bulk of high-efficiency silicon solar cells. Traditionally, defect parameters are extracted using complex techniques and equations, but the ML alternative, trained on a vast dataset of simulated lifetime curves, was shown to achieve high accuracy. Experimental validation shows that the ML predictions fall within the traditional method's uncertainty range. They argue that such advanced data-driven models can be applied to materials beyond silicon. Meftahi et al. [2] pioneered an application of ML to identify promising materials for cost-effective and printable organic solar cells (OPV). Their approach involved the integration of ML with density functional theory (DFT) calculations, enabling rapid screening and the establishment of quantitative relationships.

These relationships facilitated the prediction of critical OPV properties, including power conversion efficiency (PCE), open-circuit potential (V_{oc}), short-circuit density (J_{sc}), and highest occupied, lowest unoccupied molecular orbital i.e. HOMO and LUMO energies, and their effective band gap. Notably, their ML model exhibited remarkable predictive accuracy, achieving a precision of $\pm 0.5\%$ for both the training and test datasets. This innovative approach holds significant promise for accelerating the identification of optimal materials for OPV, offering potential benefits for affordable and sustainable solar energy solutions.

One of the issues regarding contemporary ML applications is their need to relearn tasks. Kaya and Hajimirza [3] introduced a framework that allows the transfer of learning across tasks when the specifications change. Barkhordari et al. [4] analyzed the I-V characteristics of various Schottky-type structures with varying nanocomposite interfacial layers using Gaussian Process Regression (GPR), Kernel Ridge Regression (KRR), Support Vector Regression (SVR), and Artificial Neural Network (ANN). The training data was obtained from the TE method.

The ML algorithms mostly predicted accurately, with SVR performing particularly well. Similarly, Ocaya et al. recently used ANN models to analyze and extract internal parameters of Schottky photodiodes (SPDs) without prior knowledge of the complex thermionic emission (TE) expression governing device current. The ANN demonstrated high accuracy without relying on the TE equation [5].

8.2 The TE Equation in the ML Context

The TE model in Eq. 2.26 depends on ϕ, n, R_s, on the applied voltage bias V, and on the ambient parameters i.e. the absolute device temperature T, and illumination, P. An empirical data point in a typical SPD measurement (at a given P and T) consists of the external, observable diode current I, and V.

In this chapter, we assemble, train, and apply ML to evaluate the internal parameters of semiconductor photodiodes (SPDs) when their current responses to illuminations are empirically known. Three machine learning models were used to analyze the empirical datasets and deduce the internal parameters of a p-Si/Au Schottky Photodiode (SPD). Notably, we achieve this without the need to explicitly introduce Eq. 2.26 to the models. The models used are an Artificial Neural Network (ANN), Logistic Regression (LR), and Decision Tree (DT).

The Cheung-Cheung method estimates of R_s, n, and ϕ were used to assemble training, test, and validation datasets on actual SPDs, with $A^* = 32\,\text{A/K}^2\text{cm}^2$, $A = 1\,\text{mm}$, and $T = 300\,\text{K}$ [10]. The training and evaluation of the ML models were done on twelve private experimental datasets using Google Colaboratory [12]. They contain responses to the illumination falling on three photodiodes doped with 0, 1, and 10% graphene doping of the p-type silicon substrate material. Each SPD was subjected to dark up to 30 mW/cm^2 illumination. The datasets also contain the calculated values of the following parameters for each diode: barrier height (ϕ), ideality factor (n), and the series resistance (R_s).

8.2.1 The ANN Model

An ANN is inspired by the human brain and simulates an interconnected network of neurons. The depiction Fig. 8.1 shows ANN as comprising an input layer, one or more hidden layers that extract input and output features and patterns, and an output layer that produces the final classification. The output is computed using weights w_i on inputs X_i according to

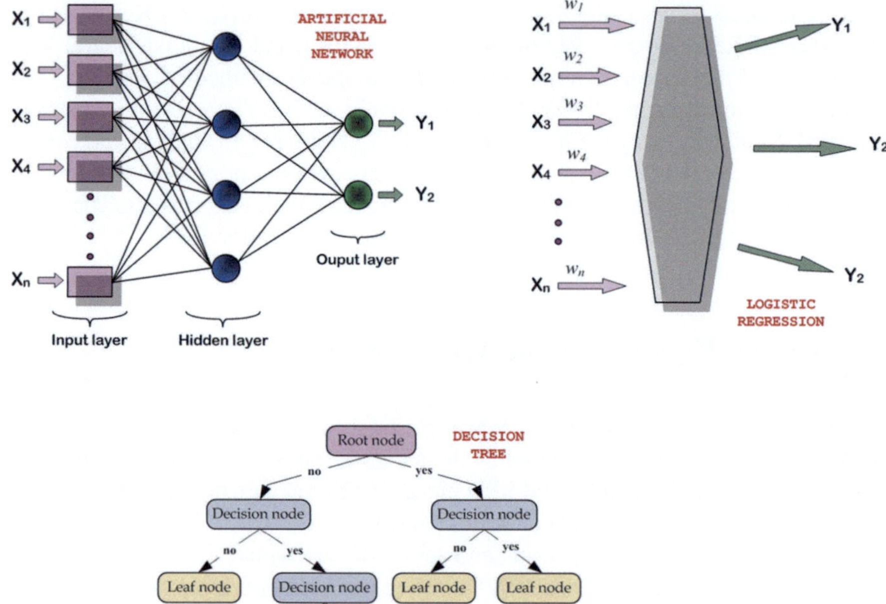

Fig. 8.1 Depictions of the ANN, LR, and DT machine-learning models used here

$$\text{Output} = \sum_{i=1}^{n} w_i X_i + b. \tag{8.1}$$

A bias factor, b, is included for flexibility in training the model. The weighted sum is input to the efficient, nonlinearity-inducing mathematical function called the rectified linear unit (ReLU) to determine whether or not a node with critical inputs is activated.

8.2.2 The LR Model

LR is a reliable algorithm for classification problems that works for binary, linear, and multi-class classification. As illustrated in the figure, its main goal is to establish the connection between dependent and independent variables by calculating probabilities through a logistic function,

$$f(x) = w^T x, \tag{8.2}$$

where w and x are the weights and inputs, respectively. The predictions of "true" are made on the output above a threshold of 0.5, and "false", below.

8.2.3 The DT Model

The DT model is a supervised ML algorithm for classification and problems that involve regression. It is a tree-like structure with root, internal, and leaf nodes that uses a divide-and-conquer approach, finding optimal split points for each node through a greedy search from the root. An example of a DT is the random forest (RF) algorithm [1]. Information gain is calculated using algorithms such as Gini and Entropy to select the best-split points for each node, which are defined as

$$\text{Gini} = 1 - \sum_{i=1}^{n} p^2(c_i), \tag{8.3}$$

$$\text{Entropy} = \sum_{i=1}^{n} p(c_i) - p(c_i) \ln(p(c_i)), \tag{8.4}$$

where $p(c_i)$ is the probability that class c_i belongs to a node. The information gain (IG) is then the entropy of the parent node minus the average entropies of the children nodes. DTs recursively split nodes based on the best-split point for each node until most or all data points are classified, with the level of classification depending on the complexity of the tree.

8.3   Dataset Creation

The datasets presented to all the developed ML models are based on V and I data points measured on Al/GO:CoPc/p-Si/Au diodes [10], together with the instantaneous estimates of ϕ, n, and R_s. Table 8.1 shows the known results of the Cheung-Cheung method at 0, 1, and 10% GO content [10].

Each dataset entry therefore has the structure (V, I, ϕ, n, R_s) for each illumination intensity. The four illumination intensities used are 0 mW/cm^2 (dark), 10 mW/cm^2, 30 mW/cm^2, and 60 mW/cm^2. Three private datasets were collected, denoting 0%, 1%, and 10% GO, respectively. Figure 8.2 shows the plot of the raw empirical data collected for these diodes over the illumination range from dark to 100 mW/cm^2. Only the 0 to 60 mW/cm^2 intensities were used in the development of the models. The 80 and 100 mW/cm^2 intensities were used as data for prediction during the ML

Table 8.1 Empirical results using the Cheung-Cheung functions for the Al/GO:CoPc/p-Si/Au diodes [10]. The entries at 80 and 100 mW/cm^2 show predictions based on the dataset during model development

Intensity (mW/cm^2)	0% GO			1% GO			10% GO		
	n	Rs (kΩ)	ϕ (eV)	n	Rs (kΩ)	ϕ (eV)	n	Rs (kΩ)	ϕ (eV)
0	13.28	0.516	0.583	13.55	4.42	0.737	22.44	4.32	0.707
10	23.78	0.509	0.579	12.44	4.90	0.764	20.91	4.49	0.686
30	27.15	0.468	0.563	12.85	4.60	0.757	6.60	4.13	0.874
60	23.00	0.516	0.577	9.89	6.60	0.807	10.86	2.07	0.758
80	23.85	0.497	0.572	12.11	4.90	0.764	11.16	1.58	0.753
100	22.23	0.502	0.576	11.00	4.54	0.776	12.39	1.72	0.736

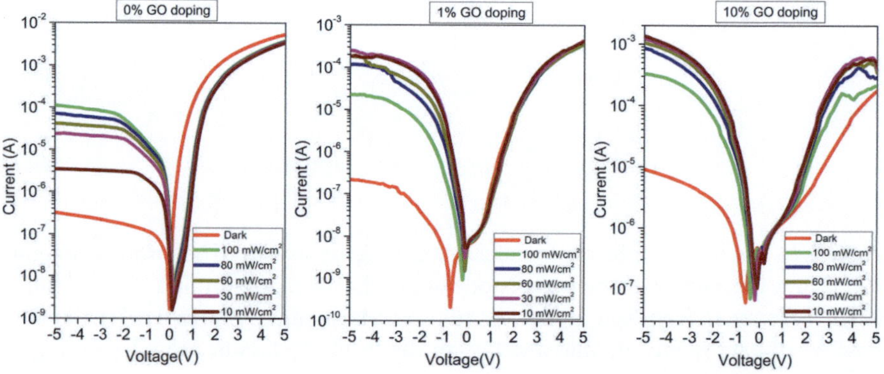

Fig. 8.2 The plots of the measured current-voltage characteristics of the variously doped Al/GO:CoPc/p-Si/Au diodes over the illumination intensity range of 0 to 100 mW/cm^2

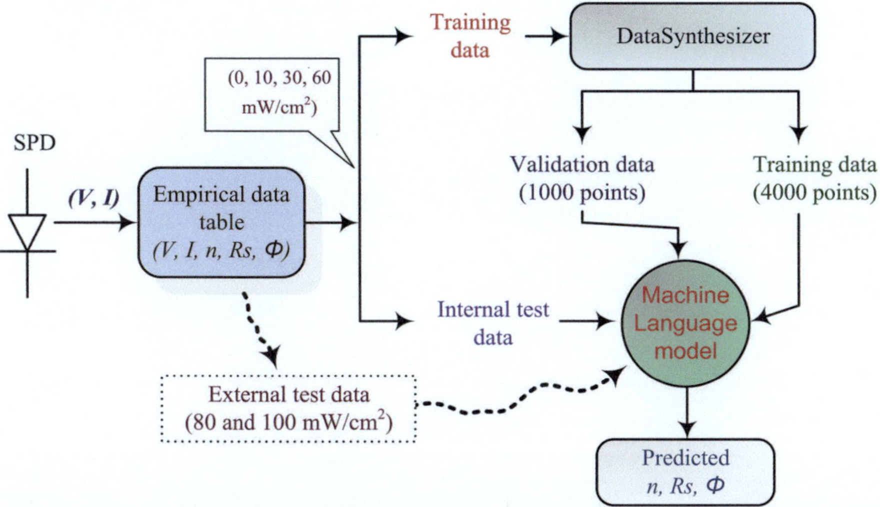

Fig. 8.3 A block schematic representation of the machine language models and the data flows in the processing. The Machine Language block denotes either ANN, LR, or DT models

model development. In all, there are 12 datasets, and each of them consists of 101 sample points along the current-voltage characteristics, from -5 V to $+5$ V, including 0, in steps of 0.1 V.

The random forest ML modeling approach for small datasets [11] by generating 5000 structurally and statistically similar synthetic Gaussian samples, shown in Fig. 8.4, for each subset using the DataSynthesizer tool [8]. The datasets were standardized before training, and also contain for each diode, the calculated ϕ, n, and R_s. Standardization was done to ensure that all the responses in the dataset contribute equally to the applied ML model. Three ML algorithms were developed and trained on the twelve datasets to model the photo responses, namely Artificial Neural Network (ANN), Logistic Regression (LR), and Decision Tree (DT). As shown in Fig. 8.1, the final ANN model has three layers: one input, one hidden layer, and one output consisting of three neurons, one each for ϕ, n, and R_s. The output from the first and second layers is passed through dropout, over-fitting protection layers. The dropout rates used for all the models are shown in Table 8.2. One LR and one DT model were developed for each dataset. The Adam optimizer [17] was applied to all the models.

8.3.1 Training and Testing

The same learning rate (0.001), L2 regularizer (0.001), and kernel constant (3) were used for all the models. They were each trained on 4000 synthetic samples and validated on 1000 synthetic samples. Each dataset was used to train a separate model,

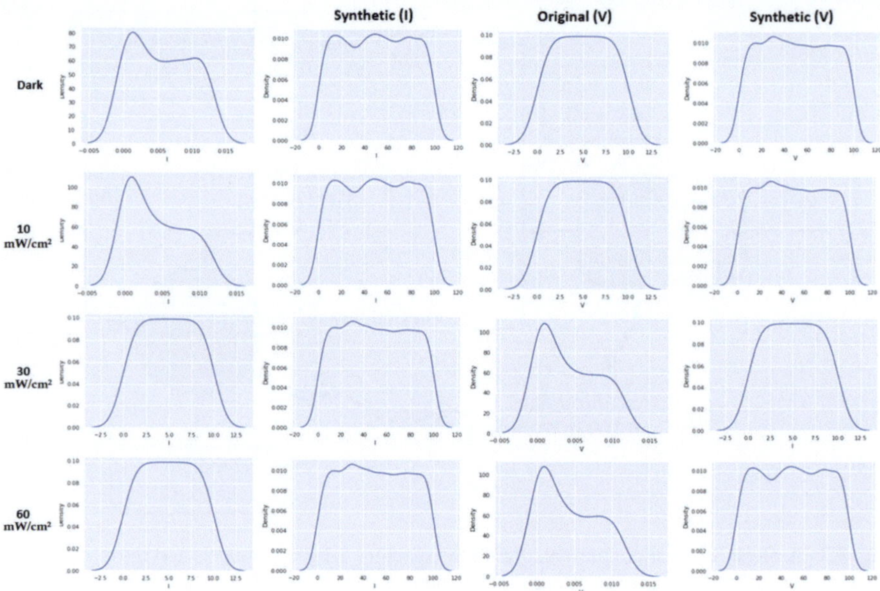

Fig. 8.4 The original and synthetic datasets have a normal distribution which is inclined to ML modeling

Table 8.2 The training parameters for the ANN machine language model

GO doping	Illumination (mW/cm^2)	Batch size	Dropout rate	Epoch #
0%	0	64	0.3	15
	10	32	0.2	15
	30	32	0.2	15
	60	32	0.2	15
1%	0	32	0.2	15
	10	64	0.4	15
	30	64	0.4	15
	60	16	0.3	15
10%	0	16	0.4	15
	10	16	0.4	15
	30	16	0.4	15
	60	32	0.3	15

and during the training process, two independent variables (V and I) and three target variables (ϕ, n, and R_s) were used as inputs. The ANN model was trained on twelve different datasets using the parameters in Table 8.2. The training relied on 4000 samples from synthetic datasets and was validated on 1000 samples.

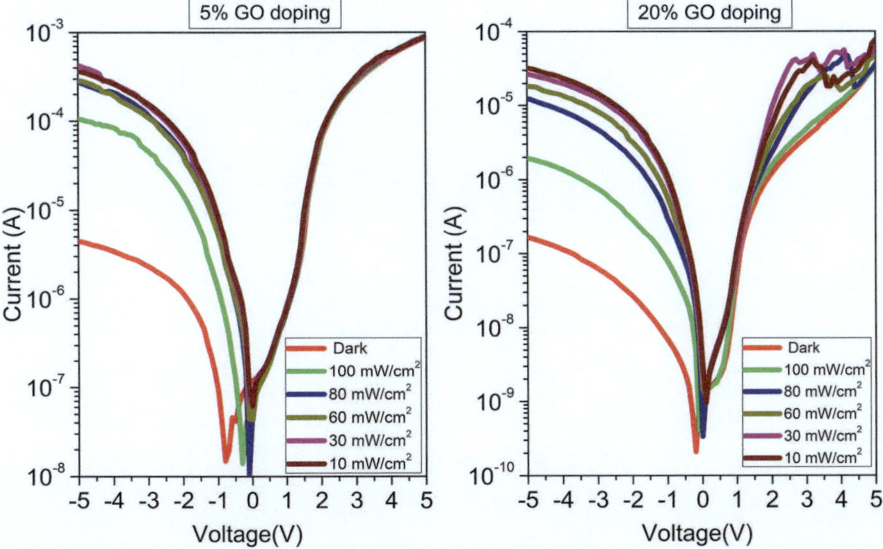

Fig. 8.5 Plots of additional original datasets at 5% and 20% GO doping levels, which were not used in the ML model development stages

Finally, the trained model was tested on all samples from the original datasets. The trained models were then tested on the 101 original samples. Tables 8.3 and 8.4 also show the validation and test results, together with the mean squared error (MSE), mean averaged error (MAE), and root mean squared error (RMSE) for all the ML models. Both LR and DT models correctly predicted all the values of ϕ, n, and R_s, indicated by a zero value of each of MSE, MAE, and RMSE. Figure 8.5 plots the current-voltage characteristics for external data that were used to further validate the ML models.

8.4 Typical ML Results

8.4.1 ANN Model

Table 8.3 shows the train and validation accuracy for the ANN model after 15 Epochs. An Epoch marks the processing of all data once. It typically involves several iterations, which can involve data batches of a specified size. Additionally, the table provides the average test accuracy for each model, which was calculated by summing the three test accuracies and dividing them by three. These findings suggest that the ANN model achieved good accuracy in predicting the target variables. The results also indicate that the model's performance varied depending on the dataset used for

Table 8.3 Training, validation, and MSE accuracy results for ϕ, n, and R_s using the developed ANN machine language model for SPD characterization on the experimental dataset

GO doping (%)	Dataset	Train Acc	Val Acc	Test Acc					MSE		
				n	R_s (kΩ)	ϕ (eV)	Average		n	R_s (kΩ)	ϕ (eV)
0%	0	100	100	99.99	99.99	99.99	99.99	3.91E-07	2.72E-08	4.29E-09
	10	100	100	99.99	99.99	99.99	99.99	6.87E-07	3.34E-09	3.81E-09
	30	100	100	99.99	99.99	99.99	99.99	3.81E-07	5.25E-09	2.34E-08
	60	98.12	100	100	99.99	99.99	99.99	0.00E+00	2.72E-08	2.19E-08
1%	0	100	100	99.99	99.99	99.99	99.99	1.91E-07	7.63E-08	1.14E-08
	10	100	100	99.95	99.95	97.27	99.059	5.66E-03	2.51E-03	2.08E-02
	30	100	100	99.99	99.99	99.99	99.99	3.81E-07	9.54E-08	2.91E-08
	60	100	100	99.99	99.99	99.99	99.99	3.43E-07	9.54E-08	1.86E-08
10%	0	100	100	99.999	99.999	99.999	99.999	5.34E-07	1.72E-07	1.72E-08
	10	100	100	99.981	99.40627	98.43502	99.27417	3.92E-03	2.67E-02	1.07E-02
	30	100	100	99.9999	99.99999	99.99999	99.99999	9.54E-08	1.14E-07	1.29E-08
	60	100	100	99.99999	99.99999	99.99999	99.99999	3.43E-07	6.68E-08	1.62E-08

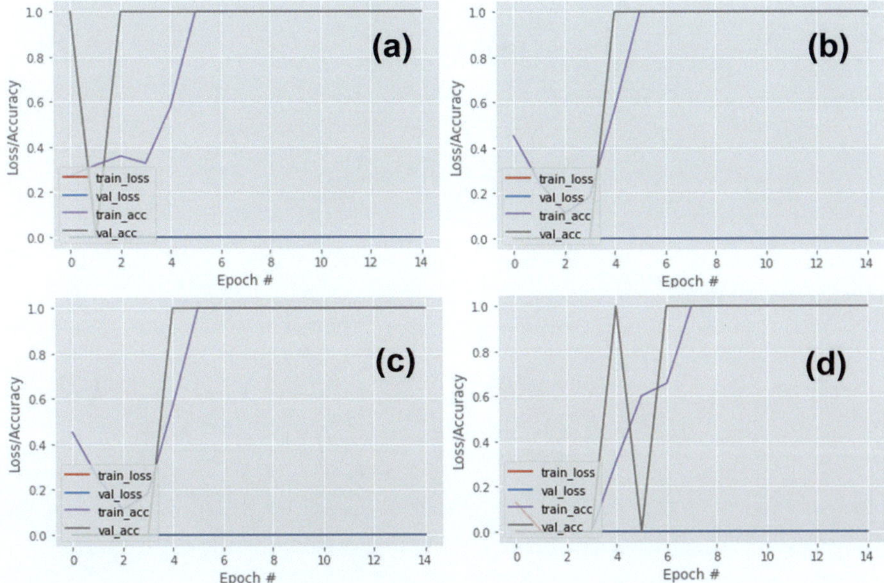

Fig. 8.6 Plots showing the training and validation loss and accuracy for ANN model at 0% GO doping. The insets **a–d** denote sequential illuminations of 0, 10, 30, and 60 mW/cm^2

training. This could be due to differences in the characteristics of the datasets, such as the range and distribution of the variables.

Further investigations are needed to determine the factors affecting the model's performance and to optimize its training parameters. In summary, the ANN model trained on twelve datasets using two independent and three target variables showed promising results in predicting ϕ, n, and R_s. Figures 8.6, 8.7 and 8.8 show the training loss and accuracy curve for some models. The plots show the model's performance for 15 Epochs. The figures show that both the train–validation accuracy curve distance, and the train–validation loss curve distance is small, indicating that there is no overfitting in the ANN models.

8.4.2 LR and DT Models

Table 8.4 shows the test and validation and test accuracies of both LR and DT models. The accuracies are 100% on all the datasets, with no overfitting.

The LR and DT models demonstrate learning of the illumination responses of the photodiodes and predict ϕ, n, and R_s with high accuracy. The table also shows the average MSE, MAE, and RMSE for all the datasets. The LR and DT models achieved a validation and test accuracy of 100%, showing that they correctly learned the light responses of the photodiodes all the values of ϕ, n, and R_s without having

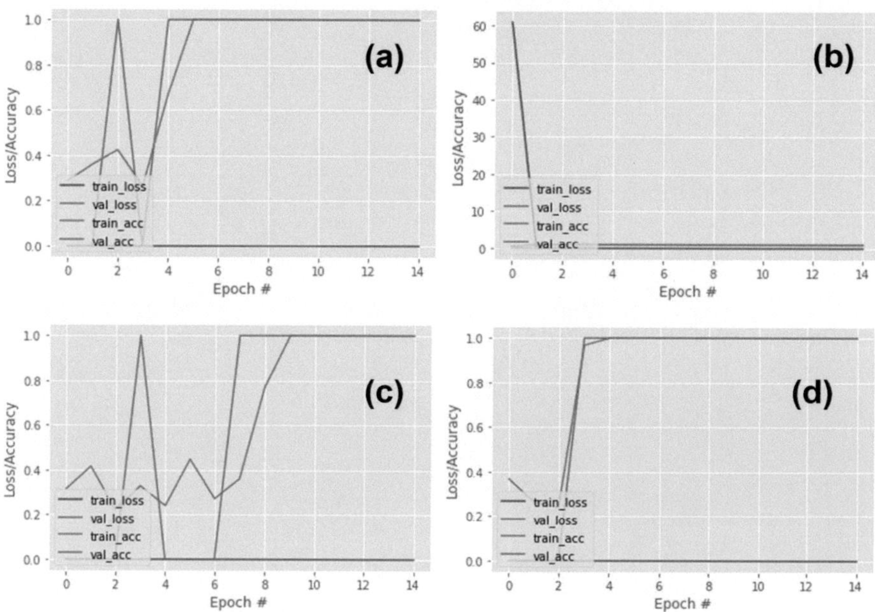

Fig. 8.7 Plots showing the training and validation loss and accuracy for ANN model at 1% GO doping. The insets **a–d** denote sequential illuminations of 0, 10, 30, and 60 mW/cm^2

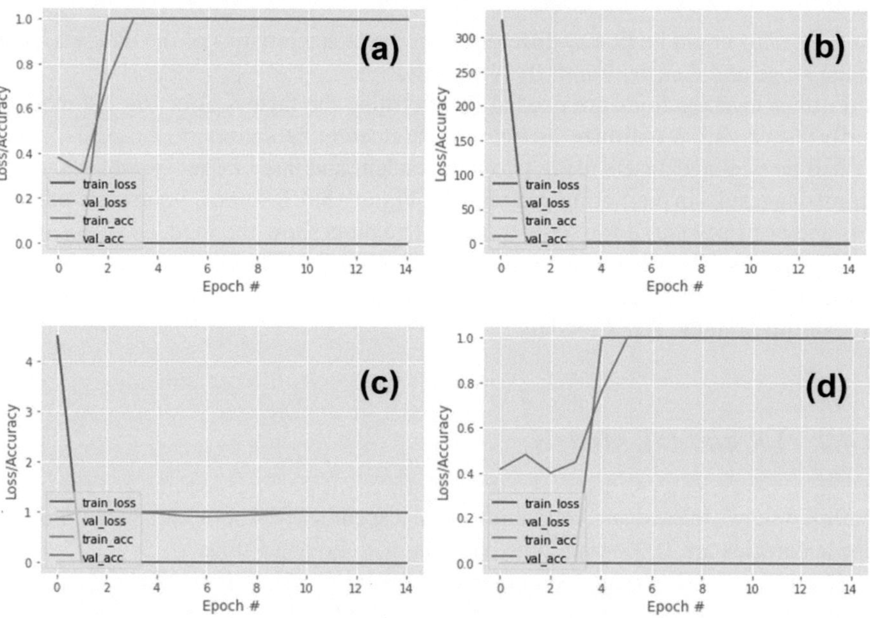

Fig. 8.8 Plots of training and validation loss and accuracy of ANN model at 10% GO doping. The insets **a–d** denote sequential illuminations of 0, 10, 30, and 60 mW/cm^2

Table 8.4 Results of the training, validation, and MSE accuracies for ϕ, n, and R_s using the developed LR and DT machine language model for SPD characterization on the experimental dataset

GO doping %	Dataset (mW/cm^2)	Val Acc	Test Acc	Test MSE	Test MAE	Test RMSE
0,1,10	0, 10, 30, 60	100	100	0	0	0

Table 8.5 Final predictions of ϕ, n, and R_s using the LR and DT machine language models for the SPD within the model development regime. The results confirm the reported values for this device [10]

GO doping (%)	Intensity (mW/cm^2)	n		R_s (kΩ)		ϕ (eV)	
		LR	DT	LR	DT	LR	DT
0%	0	13.30	13.30	0.516	0.516	0.585	0.583
	10	23.55	23.55	0.508	0.509	0.578	0.579
	30	27.25	27.30	0.468	0.466	0.565	0.563
	60	23.00	23.00	0.516	0.516	0.575	0.575
1%	0	13.55	12.44	4.450	4.430	0.735	0.736
	10	12.43	12.43	4.900	4.900	0.764	0.765
	30	0.00	0.00	4.600	4.600	0.757	0.756
	60	0.00	9.90	6.600	6.600	0.806	0.806
10%	0	0.00	22.40	4.320	4.320	0.706	0.705
	10	20.55	20.55	4.490	4.490	0.685	0.686
	30	6.60	6.60	4.130	4.150	0.865	0.873
	60	10.85	10.85	2.050	2.050	0.758	0.758

explicit knowledge of Eq. 2.26. Table 8.5 reports the performance of the LR and DT models in predicting the actual values of the internal parameters of the SPD under test [10].

Table 8.6 shows the last prediction test. It is based on the previously unspecified doping levels of 5% and 20% GO. These are shown as the external dataset in Fig. 8.3. The LR model was chosen for this test. Also shown in the table are the predicted ϕ, n, and R_s on external datasets recorded at 80 and 100 mW/cm^2.

The results show that ANN achieved a training and test accuracy of 100% and 99.99% respectively for most datasets, whereas the LR and DT models achieved a validation and test accuracy of 100%. Thus, the three ML models efficiently learned the photo responses of the photodiodes and correctly predicted their barrier height, ideality factor, and series resistance with extreme accuracy. The ML models do not need prior knowledge of the mathematical TE model in any form to reach their predictions.

Table 8.6 Final predictions of ϕ, n, and R_s using the DT ML models for the SPD with 5% and 20% GO doping levels from completely external data not part of the training regime [10]

Intensity (mW/cm^2)	5% GO			20% GO		
	n	R_s (kΩ)	ϕ (eV)	n	R_s (kΩ)	ϕ (eV)
0	10.80	2.70	0.733	26.30	1.55	0.742
10	8.32	3.03	0.793	26.33	1.62	0.733
30	8.72	2.85	0.766	26.71	1.59	0.738
60	7.53	3.06	0.813	26.85	1.77	0.733
80	4.74	3.47	0.984	19.45	2.31	0.798
100	8.09	2.74	0.781	29.94	1.47	0.706

8.5   Potential of AI in Parameter Extraction

The ANN, LR, and DT models have demonstrated the ability to deduce patterns in data pertaining to the transfer characteristics of the SPD without any prior knowledge of the device physics or the TE equation. This is achieved through training on a small set of data, allowing the ML model to generalize and make accurate predictions on unseen data. The current-voltage relationship of SPDs is governed by the highly non-linear TE equation. In the context of parameter extraction for a given SPD, it is essential to linearize specific regions of the V-I characteristic.

Many methods, like the previously described Cheung-Cheung method, are viable to accomplish this task but require a carefully selected voltage range to give accurate results. As a consequence, different individuals may select different ranges, leading to a significant variance in the extracted parameters even for a given SPD. During the implementation of the linear regression (LR) and decision tree (DT) models, it was observed that the algorithms converged rapidly and unexpectedly to the optimal ranges of applied bias across all instances.

8.6   Remarks

The above findings demonstrate the effectiveness of using ML-based techniques, specifically ANN, LR, and DT algorithms, to accurately model the light responses of photodiodes. Through extensive training and evaluation of data collected from three photodiodes, the models successfully predicted critical parameters, such as barrier height, series resistance, and ideality factor. Importantly, the models also exhibited the ability to estimate photodiode light responses under varying illuminations and voltage settings, demonstrating their broad applicability. Furthermore, the models proved capable of predicting responses with minimal error, from $0\,\mathrm{mW/cm^2}$ to $30\,\mathrm{mW/cm^2}$.

However, complete reliance on machine learning (ML) models may not provide a comprehensive understanding of peculiarities in the data, such as negative differential conductance regions or breakdowns. Moreover, only one type of Schottky barrier diode (SBD) with p-Si/Au construction were explored. However, SBDs with different constructions exhibit similar characteristics to the present devices, and it is highly likely that the same models can be used to determine their internal parameters. Further investigations are necessary to verify this, as the scope of this work is not intended to be exhaustive, but rather to demonstrate the potential of ML tools. Researchers can utilize these models to minimize the time and resources required for conducting experiments.

These findings provide an exciting avenue for future research through the application of ML techniques to model complex and highly nonlinear systems, thereby enhancing our overall understanding of their behaviors. Finally, the dataset [13] and models [14–16] are available upon request to enable independent evaluation.

References

1. Y. Buratti, Q.T. Le Gia, J. Dick, Y. Zhu, Z. Hameiri, Extracting bulk defect parameters in silicon wafers using machine learning models, NPJ. Comput. Mater. **6**(1), 142 (2020). https://doi.org/10.1038/s41524-020-00410-7
2. N. Meftahi, M. Klymenko, A.J. Christofferson, U. Bach, D.A. Winkler, S.P. Russo, Machine learning property prediction for organic photovoltaic devices. NPJ Comput. Mater. **6**(1), 166 (2020). doi:0.1038/s41524-020-00429-w
3. M. Kaya, S. Hajimirza, Using a novel transfer learning method for designing thin film solar cells with enhanced quantum efficiencies. Sci. Rep. **9**(1), 5034 (2019). https://doi.org/10.1038/s41598-019-41316-9
4. A. Barkhordari, H.R. Mashayekhi, P. Amiri, S. Özçelik, Ş Altındal, Y. Azizian-Kalandaragh, Machine learning approach for predicting electrical features of Schottky structures with graphene and $ZnTiO_3$ nanostructures doped in PVP interfacial layer. Sci. Rep. **13**(1), 13685 (2023). https://doi.org/10.1038/s41598-023-41000-z
5. R.O. Ocaya, A.A. Akinyelu, A.G. Al-Sehemi, A. Dere, A.A. Al-Ghamdi, F. Yakuphanoğlu, Machine learning models for efficient characterization of Schottky barrier photodiode internal parameters. Sci. Rep. **13**(1), 13990 (2023). https://doi.org/10.1038/s41598-023-41111-7
6. A.J. Figueredo, P.S.A. Wolf, Assortative pairing and life history strategy—a cross-cultural study. Hum. Nat. **20**, 317–330 (2009). https://doi.org/10.1007/s12110-009-9068-2
7. Z. Hao, A. AghaKouchak, N. Nakhjiri, A. Farahmand, *Global Integrated Drought Monitoring and Prediction System (GIDMaPS) Data Sets*, *Figshare* (2014). https://doi.org/10.6084/m9.figshare.853801
8. H. Ping, J. Stoyanovich, B. Howe, Datasynthesizer: privacy-preserving synthetic datasets. In: Proceedings of the 29th international conference on scientific and statistical database management, pp. 1–5 (2017). https://doi.org/10.1145/3085504.3091117
9. Z. Gu, G. Xiong, X. Fu, A.W. Mohamed, M.A. Al-Betar, H. Chen, J. Chen, Extracting accurate parameters of photovoltaic cell models via elite learning adaptive differential evolution. Energy Convers. Manag. **285**, 116994 (2023). https://doi.org/10.1016/j.enconman.2023.116994
10. R.O. Ocaya, A. Dere, H. Tuncer, A.A. Al-Ghamdi, D.C. Sari, F. Yakuphanoğlu, Synthetic Metals **209**, 164–172 (2015). https://doi.org/10.1016/j.synthmet.2015.07.016
11. L. Breiman, Random forests. Mach. Learn. **45**, 5–32 (2001). https://doi.org/10.1023/A:1010933404324

12. Google Colaboratory (2023). https://research.google.com/GoogleColaboratory/faq.html. Accessed 22 Apr. 2023
13. R.O. Ocaya, A.A. Akinyelu, A.G. Al-Sehemi, A. Dere, A.A. Al-Ghamdi, F. Yakuphanoğlu, *Dataset for Schottky Photodiodes for Machine Language Models, Figshare* (2023). https://doi.org/10.6084/m9.figshare.22679278
14. *Artificial Neural Network SBD Machine language model on Google Colaboratory* (2023). https://colab.research.google.com/drive/1iA6qP6CXS6X-mS0GW5fCmJi4i78v0rai?usp=sharing. Accessed 22 Apr. 2023
15. Linear Regression SBD Machine language model on Google Colaboratory (2023). https://colab.research.google.com/drive/1nLPHjyX-TpOXnhXOls3A4N7D-rhdBmK1?usp=sharing. Accessed 22 Apr. 2023
16. Decision Tree SBD Machine language model on Google Colaboratory (2023). https://colab.research.google.com/drive/1NAIl-xTAbijJB84SYVG3Uh16jybEon30?usp=sharing. Accessed 22 Apr. 2023
17. D.P. Kingma, J. Ba, *Adam: A Method for Stochastic Optimization* (2017). arXiv:1412.6980, https://doi.org/10.48550/arXiv.1412.6980

Appendix
Evidence of Temperature Symmetry in Semiconductor Devices

Abstract The temperature behavior remains one of the most important fundamental properties of any given system that is linked inextricably to the other parameters of the system. However, despite this importance, it remains the least framed and explored properties of most systems for many reasons. The effects of temperature constrain most analyses to a constant or narrow domain where its impact on the other parameters is considered to be minimal. Even in systems where an analytical form may be available, the parametric expression of the system may possess an overall complexity that necessitates a reduced functional domain, thereby limiting the usable range of the analysis. Nowhere is this more apparent than in the case of semiconducting junction devices with energy barriers that are complex functions of temperature, current density, and bias. In previous work, it was shown that at a given temperature, the current-voltage characteristics have translational symmetry despite the apparent complexity of the expression. In this chapter, we extend the use of group theory in the non-linear analysis of temperature symmetry by examining the case of a simple p-n junction device. The results indicate the existence of a temperature-dependent symmetry and provide a new direction for the wide-domain analysis of temperature effects in a given system.

A.1 Introduction

Temperature is a ubiquitous and inextricable part of all physical systems. In spite of its importance, it tends to be the least quantified. The usual approach to model the temperature behaviors parameterizes it, thereby allowing an isothermal analysis to be done. This resulting simplified analysis is tedious, largely to avoid a pedantic and complex analysis toward a speedier understanding of the aggregated system response. The penalty is a limited and discontinuous analysis scope. Clearly, the overall response of a real system is the aggregate of all the individual parametric responses.

R. Ocaya, *Extraction of Semiconductor Diode Parameters*,
https://doi.org/10.1007/978-3-031-48847-4

Metal-semiconductor, or Schottky, devices have two junctions, one that is ohmic and the other rectifying. Such devices are at the forefront of modern electronics and conform to the thermionic emission (TE) equation that relates device current (I) to the applied voltage (V) [1, 2]. The TE is complex due to its many parameters and the cyclical expression of its series resistance-dependent forward current. This accounts for its notoriety as a highly difficult equation to solve analytically [3–6]. In recent work, the existence of a series resistance-dependent symmetry was proved [7], which showed that under constant temperature, the series resistance is related simply to the instantaneous derivative of the I-V characteristic.

In this chapter, we examine the potential to apply the theory of Lie groups [8] to investigate the non-isothermal, continuous temperature behavior under a constrained temperature range that limits the impact on the other device parameters. Although the method is being applied to a real p-n junction diode, the 1n4148, the general approach may be applicable to other devices. Most methods that are employed to analyze the closed-form equation of the current versus applied voltage make simplifying approximations to the equation and consequently lose a generality, whose impact is often not always fully characterized. The absence of approximations in the method being presented is a clear advantage.

A.2   Theory

In the TE theory, the p-n diode current I is given by [3, 9, 10]

$$ I = AT^{3/2} \exp\left(-\frac{E_g}{kT}\right)\left[\exp\left(\frac{qV}{nkT}\right) - 1\right], \tag{A.1} $$

where V = applied bias, T = absolute temperature, ϕ = bandgap energy, n = ideality factor, q = electron charge, k = Boltzmann constant, and A = constant. If we set $T = x$, $I = y$, $-1/k = b$, $q/nK = a$, then Eq. (A.1) can be normalized with $A = 1$ to

$$ y = x^{\frac{3}{2}} e^{\frac{g}{x}}\left(e^{\frac{d}{x}} - 1\right), \tag{A.2} $$

where $g = bE_g$ and $d = aV$.

The necessary condition for the existence of symmetry is the presence of a set of unique solutions that arise from a continuous differentiation action such that each solution still defines a valid device characteristic. This is termed closure [8, 11], such that new solutions $\hat{y}'$ can be found under differentiation:

$$ \omega(x, y) = \frac{dy}{dx} = y' \equiv \omega(\hat{x}, \hat{y}) = \frac{d\hat{y}}{d\hat{x}} = \hat{y}'. \tag{A.3} $$

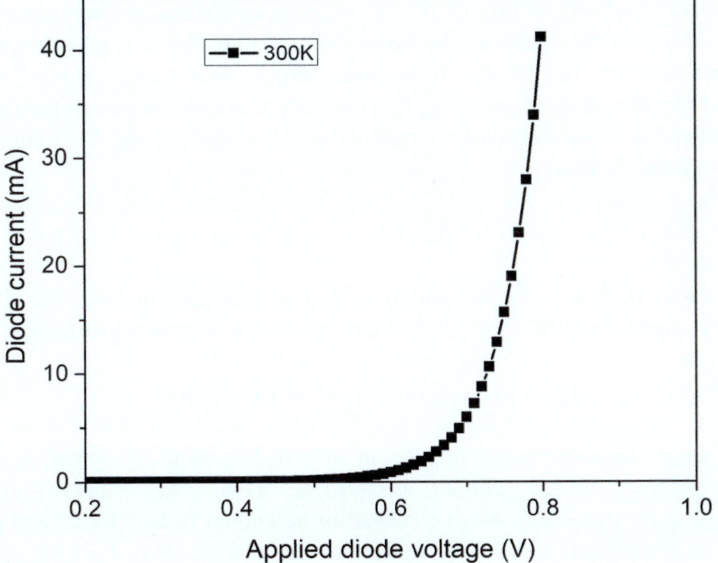

Fig. A.1 The experimental I-V plot for the 1n4148 commercial diode at 300 K. Only one temperature is plotted here in order to fit the graph on the axes shown

Figure A.1 plots the actual data from an experiment using the 1n4148 commercial diode at 300 K. that suggests the existence of a temperature-dependent symmetry in the p-n junction characteristics, all other parameters being constant.

Differentiating Eq. (A.2) w.r.t x gives

$$\omega(x, y) = y' = -d\frac{e^{(d+g)/x}}{x^{1/2}} - g\frac{y}{x^2} + \frac{3y}{2x}. \tag{A.4}$$

Setting $(d + g)$ as h, gives

$$\omega(x, y) = y' = -d\frac{e^{h/x}}{x^{1/2}} - g\frac{y}{x^2} + \frac{3y}{2x}. \tag{A.5}$$

From Eq. (A.2),

$$e^{h/x} = e^{(g+d)/x} = \frac{y + x^{1/2}e^{g/x}}{x^{3/2}}. \tag{A.6}$$

Substituting Eq. (A.6) into Eq. (A.5) gives

$$\omega(x, y) = y' = -d\frac{e^{g/x}}{x^{1/2}} - h\frac{y}{x^2} + \frac{3y}{2x}. \tag{A.7}$$

The intuitive similarity of Eqs. (A.5) and (A.7) suggests that scaling in x (i.e., temperature) is a possible symmetry by inspection. A more rigorous proof requires that the same symmetry should arise from some tangent vector field $(\xi(x, y), \eta(x, y))$. This field has the effect of realizing near-identity transformations on the (x, y) i.e., the (T, V) plane when used with a small parameter ϵ. That is, one TE solution curve is mapped onto another:

$$\hat{x} = x + \epsilon\xi(x, y) + \mathcal{O}(\epsilon^2), \quad \text{and} \quad \hat{y} = y + \epsilon\eta(x, y) + \mathcal{O}(\epsilon^2). \tag{A.8}$$

When $\epsilon \to 0$, then $\xi = d\hat{x}/d\epsilon$ and $\eta = d\hat{y}/d\epsilon$. For such a field, the linearized symmetry condition (LSC) [8, 11, 12] can be shown to generalize for the ODE in Eq. (A.3) to:

$$\eta_x + (\eta_y - \xi_x)\omega - \xi_y\omega^2 = \xi\omega_x + \eta\omega_y, \tag{A.9}$$

Transforming the above equations to an invertible canonical system is possible [7]. The arbitrary, broadly defined ansatzes $(\xi, \eta) = (\alpha(x), \Gamma(x)y + \gamma(x))$ and $(c_1x + c_2y + c_3, c_4x + c_5y + c_6)$ can readily be shown to lead to several possible symmetries. Application of the second of the above ansatzes to Eq. (A.9) gives, for the left-hand side (LHS),

$$c_4 + (c_5 - c_1)\left\{-d\frac{e^{g/x}}{x^{1/2}} - h\frac{y}{x^2} + \frac{3y}{2x}\right\} - c_2\left\{d^2\frac{e^{2g/x}}{x} + 2dh\frac{ye^{g/x}}{x^{5/2}} - 3d\frac{ye^{g/x}}{x^{3/2}} - 3h\frac{y^2}{x^3}\right.$$
$$\left. + \quad h^2\frac{y^2}{x^4} + \left(\frac{3y}{2x}\right)^2\right\}. \tag{A.10}$$

For the right-hand side (RHS),

$$c_1dg\frac{e^{g/x}}{x^{3/2}} + \frac{c_1d}{2}\frac{e^{g/x}}{x^{1/2}} + \left(2c_1h - \frac{3c_3}{2} - c_5h\right)\frac{y}{x^2} + \left(-\frac{3c_1}{2} + \frac{3c_5}{2}\right)\frac{y}{x} + \left(c_2dg + \frac{3c_3}{d}\right)\frac{ye^{g/x}}{x^{5/2}}$$
$$+ \quad \frac{c_2d}{2}\frac{ye^{g/x}}{x^{3/2}} + 2c_2h\frac{y^2}{x^3} - \frac{3c_2}{2}\left(\frac{y}{x}\right)^2 + c_3dg\frac{e^{g/x}}{x^{5/2}}$$
$$+ \quad 2c_3\frac{y}{x^3} + \left(c_4h - \frac{3c_6}{2}\right)\frac{1}{x}$$
$$- c_6h\frac{1}{x^2}$$
$$+ \quad \frac{3c_4}{2}. \tag{A.11}$$

The symmetry that arises by comparing the terms of the LHS and the RHS of the LSC is

$$\hat{x} = x, \quad \text{and} \quad \hat{y} = e^\epsilon y \tag{A.12}$$

The scaling of the current linearly ($I \equiv y$) given by $\hat{y}$ can be appreciated by noting that

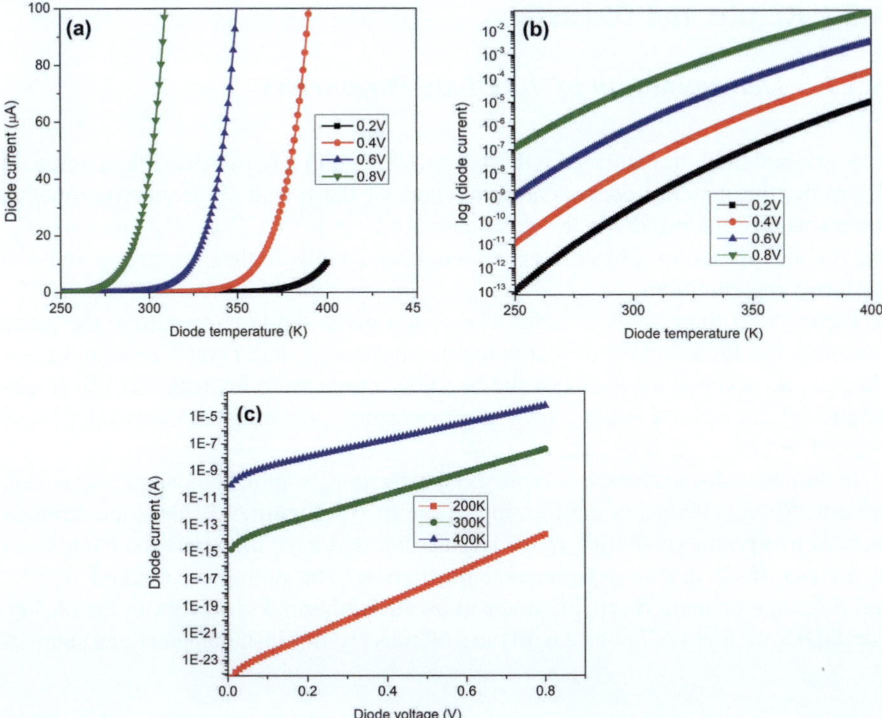

Fig. A.2 Three experimentally generated figures for the 1n4148 diode that suggest intuitively the existence of a temperature-dependent symmetry. In **a**, current-temperature plot at various applied biases. In **b**, the first plot is shown on a log-linear scale for the entire data set. In **c**, logarithm-linear plot of the diode current versus applied bias at different temperatures

$$\hat{y} = ye^{\epsilon} = (1 + \epsilon + \frac{\epsilon^2}{2!} + \frac{\epsilon^3}{3!} + \cdots) = y(1 + \epsilon + \mathcal{O}(\epsilon^2))$$

$$\approx y(1 + \epsilon)$$

$$\implies \hat{y} = \alpha y, \quad \text{as } \epsilon \to 0, \tag{A.13}$$

for some scaling constant α. This is understood to mean that one current is mapped to another solution by continuously varying the small parameter ϵ, hence meeting the symmetry condition and confirming Fig. A.2. Under this symmetry, the ODE in Eq. (A.7) is transformed to a new one,

$$\hat{\omega}(\hat{x}, \hat{y}) = \hat{y}' = -d\frac{e^{g/\hat{x}}}{\hat{x}^{1/2}} - h\frac{\hat{y}}{\hat{x}^2} + \frac{3\hat{y}}{2\hat{x}}. \tag{A.14}$$

There may be other possible symmetries but Eqs. (A.12) and (A.13) are of interest in the present scope since they define the isothermal symmetry, $\hat{T} = T$.

A.3   Results and Discussions

A.3.1   *Determination of the Diode Parameters*

The original and transformed ODEs Eqs. (A.7) and (A.14) provide a route to determine the material-dependent parameters of the p-n diode from experimental isotherms. For the p-n diode the main parameters in Eq. (A.1) are the bandgap, E_g, and the ideality factor. They are expressed, respectively, in the constants g and a in the foregoing equations.

Figure A.2b shows a clear trend where, at a given diode temperature, the diode current scales linearly over the entire temperature range under consideration. In previous work, some symmetries of the Schottky diode were investigated [7]. It was mentioned that several factors, such as temperature, impact the electrical characteristics of semiconductor devices.

In the above discussions, it was proved that the temperature behavior does, indeed, present strong evidence of conforming to symmetry. Figure A.3 shows the arrangement of two points on the I-T characteristic that was used to extract the parameters of the p-n diode in this experimental illustration. The points are marked A(x, y) and B$(\hat{x}, \hat{y})$, which are reversibly mapped by the isothermal symmetry in Eq. (A.12). The ODEs of Eqs. (A.7) and (A.14) are effectively the instantaneous gradients of

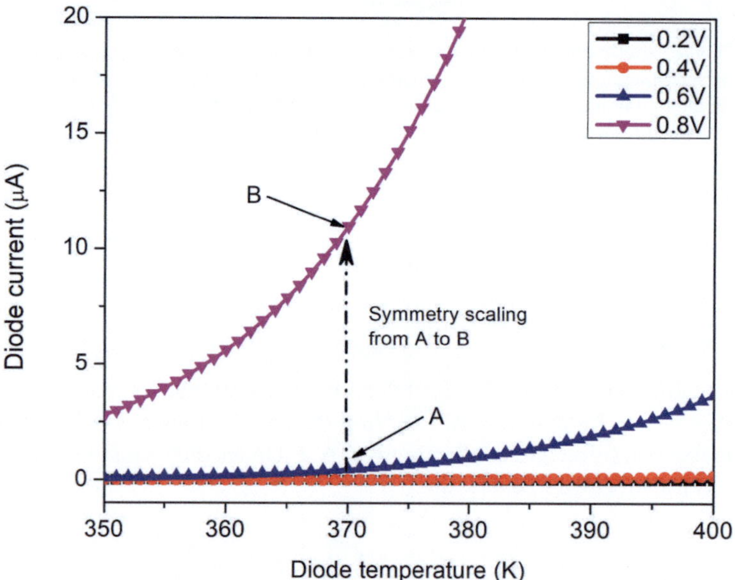

Fig. A.3 Plot of I-T for the 1n4148 showing the data points used as an illustration for the extraction of the p-n diode parameters. The dotted arrow shows the mapping of the solution at 0.6 V forward voltage to that at 0.8 V at a temperature of 370 K

Table A.1 Table of the calculated results based on the empirical measurements on the 1n4148 diode for the isotherm at 370 K. The values pertain to constants and terms in Eq. (A.6)

Parameter	Calculated value
g (eV/K)	-13003.36
d (J/K)	5805.07
n	1.98
E_g (eV)	1.07

the currents on the two curves respectively at the same temperature of 370 K. The gradients can be estimated quickly by a simple Δ method in the vicinity of the point and then applied directly to the associated ODEs i.e., $\hat{\omega}$ and ω, respectively. Then, by setting up a simple system of two simultaneous equations, while noting that $\hat{x} = x$, they can be solved for the unknowns, g and d. Table A.1 presents the results of the calculations based on the two curves shown in Fig. A.3 for the 1n4148 p-n diode. The manner in which the curves themselves are plotted reveals that they are used on a narrow region of the current, thus conforming to the condition that $\epsilon \rightarrow 0$. While the case of the p-n junction diode is presented here, one may suspect that other semiconductor devices should also exhibit symmetry. One of the advantages of following the symmetry approach is that a visual inspection of serialized curves is suggestive of some underlying relationship that betrays a symmetry. In that case, the analytical exploration through a suitable ansatz can establish the precise form of the symmetry, if any.

Although the work done so far has modeled the current along a coordinate axis of the applied voltage or, here, the temperature, one is strictly speaking not constrained by the current. Figure A.2b and c, for instance, suggest that an assignment of a function of the form $\ln(I)$ along x might also be suitable. All that remains, in that case, would then find a suitable tangent vector field and apply the LSC to it, and thus determine the underlying symmetry. For the p-n junction and other diodes, when a given output characteristic such as the forward current is kept constant such that a "serialized" set of curves are collected by repeating the experiment under different forward current conditions [13, 14].

A.4 Final Remarks

In this chapter, we have presented and explored an extension of a previously described method of symmetries for the analysis of semiconductor devices and extracting their inherent material parameters. The method also establishes possible mathematical symmetry in their characteristics through serialized responses to the typical stimuli, such as voltage and temperature. A heavy group-theoretic approach is generally avoided to keep the method simple and easier for the reader to follow. The illustration used is based on the p-n junction but is generally applicable in principle to other devices. The prerequisite is some known mathematical model for which empirical data can be fitted as proof of validity. In other words, the model should have the possibility of having a series of curves.

The advantage of the new method is that, unlike other existing approaches, the entire region of the stimulus can be analyzed. The discussion above therefore proves to be in sharp contrast to the other methods while avoiding unnecessary approximations. In general, however, methods that rely on group theory are generally considered complex and mathematically pedantic. We have shown that this need not be the case, particularly considering their overall utility on the full range of the stimuli that generate the symmetry.

Glossary

Schottky Diode   A semiconductor diode formed by the junction of a metal and a semiconductor material.

Barrier Height   The energy barrier that electrons must overcome to move from the metal into the semiconductor in a Schottky diode.

Thermionic Emission   The process by which electrons cross the Schottky barrier due to thermal energy.

Series Resistance   The total resistance in the electrical path of a Schottky diode, which affects its performance.

Space Charge Region   The region near the Schottky barrier where charge carriers are depleted.

Capacitance-Voltage (C-V) Curve   A graphical representation showing the relationship between capacitance and voltage in a Schottky diode.

Threshold Voltage   The voltage level at which a Schottky diode begins to conduct.

Majority Carriers   The type of charge carriers (electrons or holes) that dominate in a particular region of a semiconductor.

Minority Carriers   The type of charge carriers that are less abundant in a semiconductor than the majority carriers.

Reverse Bias   A condition where the voltage applied to a Schottky diode is in the opposite direction of its regular operation.

Forward Bias   A condition where the voltage applied to a Schottky diode allows current flow in the normal direction.

Breakdown Voltage   The voltage at which a Schottky diode experiences a sudden increase in current due to breakdown.

Inversion Layer   A layer of minority carriers formed near the Schottky barrier under certain bias conditions.

Oxide Capacitance   The capacitance associated with the oxide layer in a Schottky diode.

Reverse Recovery Time   The time it takes for a Schottky diode to switch off after transitioning from forward bias to reverse bias.

R. Ocaya, *Extraction of Semiconductor Diode Parameters*,
https://doi.org/10.1007/978-3-031-48847-4

Leakage Current   The small amount of current that flows in the reverse-biased condition of a Schottky diode.

Ideal Diode   A theoretical concept representing a diode with no forward voltage drop or reverse leakage.

Non-Ideal Diode   A practical diode, such as a Schottky diode, that deviates from ideal diode behavior.

Schottky Defect   A type of crystal defect that can occur in semiconductor materials.

Ohmic Contact   A type of electrical contact that exhibits linear voltage-current characteristics.

Heterojunction   A junction formed between two different semiconductor materials in a Schottky diode.

Depletion Region   The region near the Schottky junction where charge carriers are depleted due to the built-in electric field.

Schottky Equation   A mathematical expression used to describe the current-voltage characteristics of a Schottky diode.

Reverse Bias Leakage Current   The current that flows in a Schottky diode under reverse bias conditions.

Effective Barrier Height   The apparent height of the Schottky barrier as seen by charge carriers.

Doping   The intentional introduction of impurities into a semiconductor material to modify its electrical properties.

Quantum Tunneling   The phenomenon where charge carriers pass through a barrier that classical physics suggests they shouldn't be able to.

Emitter-Base Junction   The junction between the emitter and base regions in a transistor, often made using Schottky diodes.

Breakdown Voltage   The voltage at which a Schottky diode experiences a sudden increase in current due to breakdown.

Metal-Semiconductor Interface   The boundary between the metal and semiconductor materials in a Schottky diode.

Rectification   The property of a Schottky diode to allow current flow in one direction while blocking it in the other.

Ohm's Law   A fundamental law in electronics that relates voltage, current, and resistance.

Barrier Capacitance   The capacitance associated with the charge storage in the depletion region of a Schottky diode.

Avalanche Breakdown   A type of breakdown in a Schottky diode characterized by an abrupt increase in current.

Reverse Bias Leakage Current   The small current that flows in a Schottky diode under reverse bias conditions.

Ideal Diode Equation   A mathematical equation that describes the behavior of an ideal diode.

Contact Potential   The voltage difference at the metal-semiconductor interface in a Schottky diode.

Junction Capacitance   The capacitance associated with the depletion region of a Schottky diode.

Schottky Noise   The noise generated in a Schottky diode due to the random motion of charge carriers.

Barrier Height Lowering   The reduction in the effective height of the Schottky barrier under certain conditions.

Temperature Dependence   How the electrical properties of a Schottky diode change with temperature.

Barrier Width   The width of the depletion region in a Schottky diode.

Electrostatics   The study of electric charges at rest.

Diffusion Current   The current associated with the movement of charge carriers in a semiconductor.

Photovoltaic Effect   The generation of voltage and current in a Schottky diode when exposed to light.

Schottky Rectifier   A Schottky diode used in rectifier circuits.

Temperature Coefficient   How the electrical properties of a Schottky diode change with temperature.

Zener Diode   A specialized type of diode designed to operate in the breakdown region.

References

1. H. Norde, A modified forward IV plot for Schottky diodes with high series resistance. J. Appl. Phys. **50**(7), 5052–5053 (1979)
2. E.H. Nicollian, J.R. Brews, *MOS (Metal Oxide Semiconductor) Physics and Technology* (Wiley, New York, 1982)
3. E.H. Rhoderick, R.H. Williams, *Metal-Semiconductor Contacts* (Clarendon Press, 1988)
4. G. Güler, Ö. Güllü, Ş Karataş, Ö.F. Bakkaloğlu, Analysis of the series resistance and interface state densities in metal semiconductor structures. J. Phys.: IOP Conf. Ser. **153**, 012054 (2009)
5. H.E. Lapa, A. Kökce, D. Aldemir, A.F. Özdemir, Ş. Altındal, Effect of illumination on electrical parameters of Au/(P3DMTFT)/n-GaAs Schottky barrier diodes. Indian J. Phys. 1–8 (2019). https://doi.org/10.1007/s12648-019-01644-y
6. S. Khanna, S. Neeleshwar, A. Noor, Current-voltage-temperature (IVT) characteristics of Cr/4H-SiC Schottky diodes. J. Electron Dev. **9**, 382–389 (2011)
7. R.O. Ocaya, F. Yakuphanoğlu, Ocaya-Yakuphanoğlu method for series resistance extraction and compensation of Schottky diode I-V characteristics. Measurement **186**, 110105 (2021). https://doi.org/10.1016/j.measurement.2021.110105
8. N.K. Ibragimov, Group analysis of ordinary differential equations and the invariance principle in mathematical physics (for the 150th anniversary of Sophus Lie). Russ. Math. Surv. **47**(4), 89–156 (1992)
9. M. Sağlam, E. Ayyıldız, A. Gümüş, A. Türüt, H. Efeoğlu, S. Tüzemen, Series resistance calculation for the Metal-Insulator-Semiconductor Schottky barrier diodes. Appl. Phys. A **62**(3), 269–273 (1996)
10. H. Durmu, Ü. Atav, Extraction of voltage-dependent series resistance from IV characteristics of Schottky diodes. Appl. Phys. Lett. **99**(9), 093505 (2011). American Institute of Physics. https://doi.org/10.1063/1.3633116

11. P.E. Hydon, *Symmetry Methods for Differential Equations: A Beginner's Guide* (Cambridge University Press, 2000)
12. R.O. Ocaya, *Introduction to Control Systems Analysis Using Point Symmetries: An Application of Lie Symmetries* (AuthorHouse, UK, 2016)
13. A. Hussain, Temperature dependent current-voltage and photovoltaic properties of chemically prepared (p) Si/(n) Bi2S3 heterojunction. Egypt. J. Basic Appl. Sci. **3**(3), 314–321 (2016)
14. R.O. Ocaya, An experiment to profile the voltage, current and temperature behaviour of a P-N diode. Eur. J. Phys. **27**, 625 (2006). https://doi.org/10.1088/0143-0807/27/3/015

Index

R. Ocaya, *Extraction of Semiconductor Diode Parameters*,
https://doi.org/10.1007/978-3-031-48847-4

MIX
Papier aus verantwortungsvollen Quellen
Paper from responsible sources
FSC® C105338

If you have any concerns about our products,
you can contact us on
ProductSafety@springernature.com

In case Publisher is established outside the EU,
the EU authorized representative is:
Springer Nature Customer Service Center GmbH
Europaplatz 3, 69115 Heidelberg, Germany

Printed by Libri Plureos GmbH
in Hamburg, Germany